The Rise and Fall of Satellite Personal Communication Systems

Georgia Chioni

# The Rise and Fall of Satellite Personal Communication Systems

## Business and Legal Issues

PETER LANG
Frankfurt am Main · Berlin · Bern · Bruxelles · New York · Oxford · Wien

**Bibliographic Information published by Die Deutsche Bibliothek**
Die Deutsche Bibliothek lists this publication in the Deutsche Nationalbibliografie; detailed bibliographic data is available in the internet at <http://dnb.ddb.de>.

Published with financial support from the
European University Institute, Florence.

The work is based on a PhD research of
the author and expresses her personal views.

ISBN 3-631-51447-6
US-ISBN 0-8204-6488-0



www.peterlang.de

# PREFACE

**Francis Snyder**
Professor, Université d'Aix-Marseille III
Centennial Visiting Professor, London School of Economics
Professor of Law, College of Europe, Bruges and Natolin

High technology, such as satellite communication, remains a challenge, if not a mystery, to most lawyers, even more than the law does to most lay people. We read about it in the newspapers, and on the Internet, usually with little understanding how it is related to law, what is at stake, who the actors are, what potential benefits are involved and for whom, and how this new conjunction of science and business might be regulated. For non-specialists, it is not easy to appreciate how, during the past few decades, high technology and law have developed together, in ways which previously were virtually unimaginable, except in science fiction. Today, however, our lives are regulated by new forms of technology, some beneficial and others more questionable, and often governed by new types of rules, whether national regional or global in scope, public or private in origin and legally binding or not. So far, however, there are few clearly written books which could guide the lawyer or lay person through aspects of this new universe.

This book by Georgia Chioni goes a considerable way toward filling this gap in the legal literature. It is a study of the Satellite Personal Communications Systems (S PCS) market, the actors involved and the multiple legal frameworks which have shaped its development and operation. But it goes further than most law books and is really a book about law, because it analyses the business strategies of the main firms in this new market and why they succeeded or (sometimes dramatically) failed. In other words, it takes us far behind the newspapers, or the Internet, and shows how market and law are mutually constitutive and how they interact in one of the most exciting areas of current high technology. That it is also clearly written, while also dealing with legal texts and techniques, adds considerably to its value.

The work began life as a Ph.D. thesis in law at the European University Institute in Florence. It is a sign of the author's longstanding passion for high technology, and her deep theoretical and practical knowledge of the satellite communications sector and how it is regulated through law. Moreover, it testifies to her determination and success in placing this highly technical scientific material and the related complex set of legal and other norms in their wider social context. Doing law in context is not easy, and the author succeeds in her chosen task. Among the most interesting parts of the work are the discussions of relations between global markets and forms of business organization, the ways in which strategic actors, both business and regulator authorities, use the law, and the interaction between law and the market as competing or parallel types of regulation as well as explanations of (sometimes spectacular) business failure.

This interdisciplinary and multifaceted approach enables the author to escape from the Scylla and Charybdis which often trap the unwary in dealing with high technology and the law. On the one hand, it is tempting to see technology itself as autonomous, driving by itself social, economic and legal changes. On the other, it is easy to ascribe all technological development and the evolution of the legal framework to powerful strategic actors, such as transnational companies. This book avoids these traps by dealing with technology, strategic actors and the law as interconnected. A case in point is the treatment of the S PCS paradigm as a case of business failure. While the author ascribes this failure to inability to compete in the market, she also demonstrates that the market itself was embedded in society and sustained by generic European Community legislation, satellite-specific legislation and legislation specific to S PCS. This legal framework in the author's view stemmed from policy decisions and outcomes, which resulted from the interaction among strategic actors, both business and regulators.

In this specific context, the regulatory institutions and the normative framework they produced may seem especially complex. It exemplifies however a pattern of regulation which is becoming increasingly typical, in which regulation of a given subject matter, such as new technology, involves the concurrent and competing efforts of institutions at the national, regional and international as well as transnational sites of governance. Changes in these regulatory patterns and their outcome owes much, as the author of this book shows, to globalization, in the sense of the spread of models and norms from one or more sites of governance to others. One of the fascinating elements of the book is the demonstration of shifts in level of regulation, or in my terminology, shifts in sites of governance, as a result of interaction between and the relative power of strategic actors.

This book by Georgia Chioni thus contributes a good deal to educate us about how the law concerning high technology is made and applied. It also demonstrates convincingly that the method of law-in-context has much to teach us with regard to global and regional governance. It deserves to be widely read.

## Table of Contents

## List of most important abbreviations

| | |
|---|---|
| BSS | Broadband Satellite Services |
| CEPT | Administrative Conference for Post and Telecommunications |
| DBS | Direct Broadcasting satellites |
| DECT | European Cordless Digital Land-based System |
| ECTRA | European Conference for Telecommunications |
| ETSI | European Telecommunications Standardisation Institute |
| NGSOSS | Non Geostationary Orbit Satellite Systems |
| NRAs | National Regulatory Authorities |
| EEOs | Elliptical Earth Orbit Satellites |
| ERC | European Radiocommunications Committee |
| ERMES | European radio paging digital land-based system |
| ERO | European Radiocommunications Office |
| ETO | European Telecommunications Office |
| ETSI | European Telecommunicaitons Standardisation Institute |
| EUTELSAT | European Telecommunications Satellite Organisation |
| FCC | Federal Communications Commission |
| FSS | Fixed Satellite Services |
| EC | European Commission |
| EU | European Union |
| GENs | Global Economic Networks |
| GEOs | Geostationary Earth Orbit Satellites |
| GSM | Group Mobile Special |
| HEO | High Elliptical Orbit |
| ICTs | Information and Communication Technologies |
| IFRB | International Frequency Registration Board |
| IMT | International Mobile Telecommunications |
| INMARSAT | International Maritime Satellite Organisation |
| INTELSAT | International Telecommunications Satellite Organisation |
| ISLs | Intersatellite Links |
| ISOs | International Satellite Organisations |
| ITU | International Telecommunications Union |
| LEOS | Low Earth Orbit Satellites |
| MEOS | Medium Earth Orbit Satellites |
| MRC | Milestone Review Committee |
| MoU | Memorandum of Understanding |
| NCAs | National Competition Authorities |
| NRAs | National Regulatory Authorities |
| NSR | National Service Retailers |
| NSW | National Service Wholesailers |
| NTOS | National Telecommunication Organisations |

| | |
|---|---|
| PCN | Personal Communications Networks |
| PTTs | Administrations for Post and Telecommunications |
| RR | Radio Regulations |
| RSW | Regional Service Wholesalers |
| SA | Strategic Alliances |
| SNO | Satellite Network Operators |
| S PCS | Satellite Personal Communication Systems |
| SSO | Space Segment Operators |
| 3G | Third Generation |
| UMTS | Universal Mobile Telecommunications Systems |
| WARC | World Administrative Radiocommunications Conference |
| WTA | World Telecommunications Agreement |
| WTO | World Trade Organisation |

# Chapter I
# Introduction

### *I. Background*

This book examines the rise and fall of the Satellite Personal Communication Systems (SPCS) from an interdisciplinary approach. It describes a technologically advanced market formed on the basis of heavy investment, which traded international personal communication capability through the use of mobile satellite networks. Moreover, it is the story of a market where marketing miscalculations made investors pay a very high price, leading business to failure, and where co-ordinated regulatory efforts have not been enough to prevent a spectacular failure.

To undertake a market analysis is a difficult task. To undertake the analysis of a market, which had a promising beginning, and to analyse business strategies that lead pioneers to failure, is a challenge. The SPCS market is an example of this type of market, where increased and co-ordinated regulatory efforts proved unsuccessful in saving the market from collapsing. In this case, commercial needs proved that the commonly accepted rule for new markets "first comer, first winner" did not apply, as the customer base and the demand for the provided services, when the systems became operational, was much below than initially expected. Subsequent market demand and rival technologies however, showed that markets may indeed favour late comers as well.

The book aims to describe the S PCS market, the actors involved, and the legal framework that shaped its function. It aims, further, to analyse the business strategies of the economic actors in the market and present the reasons that these strategies lead to business failure. This analysis is based on the interaction of various parameters (technology, law, actors and market) forming an interesting level-playing field.

### *II. Homo Electronicus and knowledge society: economic, technological and legal challenges*

During recent decades telecommunications have been undergoing dramatic changes, which have been part of the general societal transformations in an economic, technological and legal context. From an economic point of view, phenomena such as multipolarity, transnational commercialisation, globalisation, multidimentionality of the activities and of the role of big enterprises, states, organisations and individuals have been intensified and have become major features of the post industrial era.[1] These phenomena have created a dynamic global environment, were mobility of actors has been the main feature, changing drastically the static character of interaction among players on the world stage in the previous two decades.[2] Moreover, global integration of economies and the free flow of goods and services have diminished the importance of borders, creating challenges and initiating transformations that go far beyond the power of networked relations built between national states. Policy changes such as liberalisation, deregulation and corporatisation have been some of the manifestations of the change from a state-oriented model towards a market-oriented one and an economy characterised by global economic networks (GENs)[3].

In parallel, the process of technology development has been taking place, redefining rationality, through a new mode of thinking. This mode emphasised functional relations based on criteria such as the ones of efficiency and optimisation on the basis of the rational use of resources. Furthermore, it created new economic interdependencies and new social interactions, like increasing horizontal or vertical coalitions of economic actors and establishing networked markets.[4]

Technological progress has started to change the quality and the notion of communications from a concept of limited information to a concept of hyperabun-

---

[1] See Daniel Bell, The coming of the post industrial society, a venture in social forecasting, Basic Books, Inc. Publishers, New York, 1973.

[2] Bruce R. Robert, Cunard P. Jeffrey and Director D. Mark, From telecommunications to electronic services, a global spectrum of definitions, boundary lines and structures, Butterworth, 1986, p.14. Also on the increased interaction of actors, see Peter Malanczuk, Actors: States International Organizations, Private Entities, in Lafferanderie, (ed.), Outlook on space law in the next thirty years, The Hague, 1997, p.23 ff.

[3] See Dicken Peter, Global Shift: industrial change in a turbulent world, Paul Chapman Publishing, 1988, p. 12, who argues that the bases of the global system were established much earlier, as the global system was a capitalist system, originated and developed in specific geographical locations (p.12). Also, Marc Humbert (ed.), The impact of globalisation on Euro pe's firms and industries, 1993, p.115 ff. For a general approach to the impact of globaliation of economy, see Horváth\ Merzrath, Globalisierung, Springer, 1997, p. 67 ff, Mansell Robin, Information, Organisation and Competitiveness, Networking strategies in the 1990s, in Antonelli C., (ed.) The economics of information networks, Elsevier Publishers, 1992, Francis Snyder, Globalisation and Europeanisation as friends and rivals: European Union Law in global economic networks, EUI, Working Paper, no 8, 1999, p. 9. On liberalisation, deregulation and corporatisation, see Pogorel G. and Noam E. M., Asymmetric deregulation: the dynamics of telecommunications policy in Europe and the US, Norwood: Ablex, 1994.

[4] The term "networked markets" describes markets, such as the one of telecommunications or energy, which have common structural features, being developed horizontally and vertically. See Center for Policy Research, Europe's Network Industries: conflicting priorities, Monitoring European Deregulation, CEPR, 1998, p. 5.

dance in multiple forms.[5] In this way it has signalled the beginning of a new era of communication and information, based on powerful electronic superhighways.[6]Microwaves, optical fibers, semiconductors, broadband integrated networks, personal communication networks, intelligence networks and the second generation of mobile communication systems have been major achievements of the last decades.[7]The evolution of Information and Communication Technologies (ICTs) in terms of quality and quantity of transmission, in fields such as microprocessors, hardware platforms, mobile multimedia and 3G have emerged as main features of this decade.[8]The high powered satellite technology focusing on satellite personal communications systems (S PCS) constitutes only one small part of these technological developments, which have given the push to the industry to expand and build new strategies of market identification and penetration in the satellite sector.

The specific S PCS technology emerged and started to be developed during the nineties, with the introduction of various changes within the satellite systems per se. The ability to lower the orbit from the existing geostationary one, has resulted first in an increase of the volume and quantity of transmitted data and

---

[5] Michael Marien, New communications technologies, a survey of impacts and issues, Telecommunications Policy, 20, 5, 1996; also for a presentation of electronic and digital services in the information society, see, The emerging Digital Economy, Lynn Margherio (Project Director), Dave Henry, Sandra Cooke, Sabrina Montes and Kent Hughes, US department of Commerce, 1998. On the concept of the information society, see Beniger James Ralph, The control revolution: technological and economic origins of the information society, Cambridge Harvard University Press, 1986, Deutsch Karl W., Sonntag Phillip, From the industrial society to the information society: crisis of transmission in society, politics and culture, Berlin, Wissenschaftszentrum, 1981, Forester Tom, High Technology Society: the story of the information technology revolution, Cambridge MIT Press, 1987.

[6] On the path of evolution see Braman, S., Harmonisation of systems, The third stage of information society, Journal of Communications, 1993, 43(3), 133-140. Also Public Network Europe, Getting lost on the road to the information society, October 1999. On the information society with particular emphasis on the internet, see, In Cyberspace no one can hear screeaaammming, Public Network Europe, October 1999, p. 61 ff.

[7] These networks have allowed the possibility of connection between various electronic circuits. IBN(Integrated Broadband Networks), MPN(Microwave Public Networks), PCN(Personal Communications Networks), WAN(Wide Area Networks), EDS(Electronic Data Systems-telephone systems operating through integrated GSM network), MAN(Metropolitan Area Networks), namely fiber optic based high speed networks, LAN(Large Area Networks) that interconnect ISDN with the intelligent networks.

[8] 3G(third generation) and the planned 4G(fourth generation) applications are promising integrated multimedia solutions at very high speed. On technological developments, see EITO, European Information Technological Observatory 2003, Report, 2003.

second in the multiplication of the speed of transmission.[9]As a result, S PCS satellites were seen as a means to provide services like those provided by the mobile terrestrial systems, but with the advantage of global coverage. This has made economic actors realise that these systems could provide an alternative to the existing terrestrial ones and as a result a new market could emerge which would mean new opportunities for profit maximisation.

Law has also been transformed, becoming more "flexible", in the sense that soft –non-legally binding- law documents have been multiplied in the last recent years.[10]Moreover, it has become more «global» following societal developments.[11]One of these developments has been world wide technology diffusion and creation of global markets.[12]How well can law regulate global technology and global markets?[13] It is difficult to say where the law stands: between global

---

[9] Satellites rotate the earth in various orbits. One of them -the most commonly used up to now- has been the geostationary orbit, in which the satellite rotates the earth at a height of 36.000 Km above the equator. For the technical explanations see Chapter I.

[10] On soft law, see Francis Snyder, Soft Law and Institutional Practice in the European Community, EUI, Working paper No. 4, 1993. According to Snyder, the trend of soft law represents in part a predictable feature of administrative development, in part response to institutional criteria and in part a questionable attempt to circumvent or avoid the implications of failures to reach political agreements, s. Francis Snyder, p.3, Interinstitutional Agreements: forms and constitutional limitations, EUI, Working Paper, 95/4. Also Stephen Weatherill, Law and Integration in the European Union, Oxford, Clarenton Press, 1996, p.173-174.

[11] For an extended analysis of the legal factor, see Chapters IV and V.

[12] On global law, see Francis Snyder, Global Economic Networks and Global Legal Pluralism, in George Bermann, Matthias Hedeger and Peter Linseth (ed.), Transatlantic Regulatory Cooperation, Oxford University Press, Oxford 2000, p.15, Teubner, "Global Bukowina": Legal pluralism in the world society, p.3-28, in Global law without a state, Dartmouth 1997.

[13] In the 1980s big firms started becoming aware of the fact that their competitiveness depends not only on their own strength, but also on the interaction with other forces in the environment in which they operate. The unprecedented increase in foreign direct investment, the ever increasingly intensified international rivalry and trade liberalisation, have facilitated the diffusion of technology, especially from the dominant technological power, the US. This diffusion started with evolution in microelectronics, digitalisation of switching and transmission, substitution of the electromechanical control systems with computer systems, intelligent networks, optical fibers, satellites and microwave technology. Moreover, it has contributed to the establishment of a system based on knowledge and science. When multinational firms started developing world-wide strategies, they revealed an enlarged market dimension, which has made technology spill overs unavoidable. As a consequence technology networking began to take place among the firms, which concentrated mainly of R&D. This technology networking has constituted the new phenomenon of globalisation, the one called "technoglobalism", illustrating the global diffusion of technologies, within the context of globalisation. On the impact of technology on international trade, see Raymond Vernon, Exploring the global economy: emerging issues in trade and investment, Lanham: University Press of America, 1985, Melas, Globalisation, Chapter 3. Also, Sylvia Ostry and Richard R. Nelson, Technonational-

markets and powerful networks of interaction, regulatory forces and social demand.

In order to regulate global technology and global markets, law has to have extended authority and power over all countries so as to be able to impose a global regime that would correspond to the global needs of the market in question and to legalise transborder economic relations. The question of the legal approach to technology has been under discussion. How was S PCS technology to be regulated? Two views have been expressed. One, which has been supported by the radioaviation and space industry, has argued that each technological field, particularly space technology, should be treated in a specific way. The other view, expressed by the European Union and especially the Commission, argued that satellites, like any other sector of telecommunications, should be treated without any exclusive privileges.[14]

---

ism and Technoglobalism, conflict and Cooperation, The Brookings Institution, Washington DC., 1995, p. 10, 27, Oliver Weinstein, The telecommunications equipment industry: the great transformation, Ph. Cooke, E. Moulaert, Erik Swyngedouw, O.Weinstein, P. Wells, (ed.) Towards global localisation, 1992, UCL Press, p.153, Archibugi and Michi, as above. Also Bartlett, C.A., Ghoshall, S., Managing innovation in transnational corporations, 1990, C.A. Bartlett , Y. Doz, G. Hedlund (ed.) Managing the global firms, London, Routledge, Dunning J., 1990, The globalisation of Business, London Routledge, Howells and Wood M., The Globalisation of technology, London Belhaven Press, 1993, Danielle Archibugi and Jonathan Michi, The globalisation of technology, a new taxonomy, Cambridge Journal of Economics, 1995, 19, p. 121-140, also Freeman C., and Hagedoorn J., Globalisation of technology, 1992, Brussels, EC-Fast.

[14] The Commission has been supporting a policy of technological neutrality in terms of regulation. As it has been pointed out:

"The rapid development of the sector and recent technological advances have brought about new challenges for regulators. These changes allow data to be transported through all networks, and to be accessed from a variety of terminals. This process is known as "convergence". The regulatory framework takes it into account by avoiding favouring one technology over another. This approach is called 'technological neutrality'. Technology-neutral regulation allows provision of new services and leads to competition between different access methods, known as facilities-based competition. In the mid to long-term this is the best way to low prices and increased choice of services. It also stimulates innovation and creates resilience in communications infrastructure as a whole." See Communication from the Commission to the Council, the European Parliament, the Economic and Social Committee and the Committee of the Regions, Electronic Communications: the road to the knowledge economy, COM (2003)65, 11.2.2003, p. 5.

### *III. Aim of the Book*

### 1. S PCS landscape and the context of interaction: Technology, actors, law and market

Market and business strategies are being presented in the S PCS context through the analysis of the interaction of certain variables:
*Assumption no 1:* Economic actors began building business networks, dealing with the distribution of resources, especially those concerning investment and technology. They also started interacting with regulators by campaigning for the allocation of resources, such as spectrum, and by trying to influence the setting of conditions for the market function, (entry barriers, competition, and terms of exit). As a consequence, the interplay of economic actors was mobilised by the will to get more resources, share risk and increase profits, proof of which have been the numerous partnerships focusing on S PCS projects requiring huge investment.

*Assumption no 2:* Law has been an important element of the whole interaction. It has had effects on politics, economy and society and constituted a variable, on the one hand distinct from polity and economy, but on the other interdependent, moving, in this way, to levels beyond the elaboration of the legal doctrine.[15] Legal developments can be understood by taking into account the needs of the various interest groups and their interaction with other groups of actors, in our case the interaction of economic actors with the regulators. Interactions of economic players have shown that they want to exploit resources, identify and penetrate new markets, become more efficient and initiate profitable enterprises. The S PCS market has been formed as there was demand and supply; the economic players involved were seeking regulation to govern activities, in which they were engaged, as a «safety net» mainly against lack of national administrative co-ordination and law harmonisation. They had therefore started interacting with regulators aiming to create conditions that would govern the exchange of resources and the further interaction with the customers.

*Assumption no 3:* The S PCS market emerged as a result of the interplay of the above mentioned parameters: the impact of the resource allocation on the economic actors, their interaction with the regulators, and the regulatory outcome as a result of policy output. Law shaped the characteristics of the market, but the success or failure of the product or technology of services offered has been dictated by the market itself. If a new entrant did not respect the needs and the demands of the market (e.g. customer needs, target markets, existing competitors),

---

[15] See Francis Snyder, New Directions in the European Community Law, Introduction, Weidenfeld and Nicolson, London, 1990.

then failure would be quite likely. The rise and fall of Iridium, the first operational S PCS, was an example of such a failure. This was followed by the failure of other systems as well.

*Assumption no 4*: The establishment of the market set a new context of interaction. Once the market was structured and started functioning, this structure had an impact not only on the existing actors, but also on the future actors. It was in this situation that economic actors developed their market and business strategies, which were not carefully planned, before entering the market. Analysis and interpretation of what the market can offer, is a precondition for success; overestimation and overenthusiasm, not based on solid market data may lead to business failure.

Satellite Personal Communciation Systems

| The Factors (Level 1) | Technology | Actors | Law | Market |
|---|---|---|---|---|
| Level of Interaction (Level 2) | ICTs | GENs/ SAs | Law generating/regulatory process in telecom market | Market Building |
| Theoretical basis (Level 3) | Globalisation | | Information | Society |

**2. Meaning of Concepts**

**A. Resources**

Economic relations are defined, driven and characterised through the exchange of resources. The type of resources exchanged differs depending on the actors that participate, as different groups of actors possess distinct kinds of resources.For instance, multinationals possess wealth, regulatory bodies may offer legitimacy and states their competence to monitor agreements.[16]The term "resources" has a wide meaning. Resources are elements of great importance for the strategic actors and their distribution may influence and determine power relations among them. A «resource» may be anything that can be exchanged and which is advantageous to the owners, such as human resources, know- how or

[16] See Stephen D. Krasner, Power politics, institutions and transnational relations in Thomas Risse-Kappen (ed.), Bringing transnational relations back in, Non-state actors, domestic structures and international institutions, Cambridge University Press, 1995.

money.[17]In particular, it has been defined as "[....] a good (material or immaterial), a value, whose allocation may determine the position of the individual important resources." In this study the most precious resources are considered to be technology, spectrum and investment.[18]

The most difficult issue related to resources has been the one of access, which determines first, whether the actors will be able to enter the market and second, whether the actors that have already entered the market will be able to survive competition. The resources analysed below - technology, spectrum and money- are necessary pre-requisites for the entry of the market players and remain decisive factors for their success in the market.

a. Technology

Technology is a resource that has been improving human living conditions in various ways. For instance, it has increased the number of goods and the pace of mass production as well as reduced prices. Moreover, it has contributed to the creation of new classes of technological specialists, engineers and technicians, who are important for the creation and operation of technological systems.

Technology can be everything, form railways to computers, energy and satellites. For each of the actors that come into contact with the technology, the meaning changes. For instance, for the economists, technology is everything that makes production function better. For politicians, technology is a means of increasing power, for consumers, everything that makes lives more comfortable.[19]

---

[17] See Fawcett J. E. S. and Audrey Parry, Law and international resource conflicts, Clarenton Press, Oxford, 1981, p.1.

[18] The authors Fawcett and Parry attempt a general definition of resources, where resources "may be taken to comprise both the source materials for energy, food, manufactures and the means of obtaining them, particularly industrial and agricultural know how, capital investment, foreign exchange and transport services. Source materials have various overlapping characteristics: as obtainable by extraction, as durable or perishable; or as primary or semi-processed products. Some may be renewable, some may be in the course of development." Fawcett/Parry, Law and International Resource Conflict, as supra.

[19] James B. McOmber attempts an interesting theoretical approach, by giving a very broad definition and tracing three functions. First, technology as instrumentality, in which case autonomy is the most representative feature. Second, technology as industrialisation, where technology is presented as a product of a specific time and frame. And third, technology as novelty, where it is presented as a narrative of continued discontinuity. In all cases, technology has as a major feature autonomy. See James McOmber, Technological autonomy and the three definitions of technology, Journal of Communications, summer 1999, v.49, n.3. On technology as an autonomous concept, which initiates societal evolution, see Ellul J., 1964, The technological society, NY: Vintage, Feenberg A., 1991, Critical theory of technology, Oxford, UK, Oxford University Press. Technology as a generic concept forms an autonomous theoretical entity. However, when one analyses societal transformations, examining technolo-

When new technology is used by the industry, it is either adopted by the existing firms and there is no significant change, or there are significant changes in the industry.[20]There have been technological changes that have affected industrial structures. One of these has been the development of satellite technology, which has emerged as a cost-effective method of transmitting communications.[21]This technology, being used and exchanged in multiple forms in the recent decades, has given the push to the industry to expand and build new strategies of market identification and penetration in the satellite sector.

b. Spectrum

Spectrum constitutes the second important resource, as it has been the decisive factor for all players that would to enter the market.[22]For them spectrum has been the means to become competitive and claim market share. For regulators allocating and assigning frequencies has been a difficult task because of highly complicated technical particularities and political and economical implications. For the states, spectrum has been related to sovereignty and economic gains.[23] Spectrum, the "common heritage of the Mankind", as it has been characterised,[24]

---

gy within the societal environment, it is difficult to consider it as an autonomous entity, as it gets integrated into the societal environment.

[20] Gerald W. Brock, The telecommunications industry, the dynamics of market structure, Harvard University Press, 1981, p.22.

[21] Ibid, p. 245.

[22] On technical explanations with regards to spectrum, see Chapter II.

[23] Spectrum is a definite natural resource and states, being territorial rulers, have been the entities responsible for spectrum assignment to those interested parties that want to provide services in a certain country.

[24] This characterization has become one the principles of international law, reaffirmed in the Outer Space Treaty (1967) concerning space, geostationary orbit, spectrum and celestial bodies. The initial elaboration of the notion belongs to O.Schachter (O.Schachter, Who owns the Universe? US Symposium 1961, p.844, 846), mainly concentrating on the fact that the celestial bodies were considered to be the common property of all states. The concept "common heritage of the Mankind" as such, was introduced for the first time by the UN-Ambassador of Malta, Arvid Pardo, in October 1987, at a UN-general meeting concerning the new law of the sea. The concept of a legal state element was negatively repealed. The exploration should be equal for all states. The term has in the meantime been internationally established and accepted. According to Wolfurm (The principle of Common Heritage of the Mankind, ZaöRV, 1983, 312), the elements which a good has to have so as to be characterised as a common heritage of the mankind are the following. First, the good must belong to the whole of mankind that must leave it free for utilisation. Second, the nations or the international organisations act as in the name of the mankind, when they use it. Third, all states must have equal access to its utilisation, with no discrimination. Its characterization as a common heritage has initiated endless debates regarding its nature: what it is and its limits, in other words: where does air end and where does space begin. More than 35 theories have been developed trying to deal with these questions. See for an analytical presentation, Mateesco Mattee, Space Law, in Encyclopedia of public international Law, v. 11, Law of the Sea, Air and Space, 1989, 303-

constitutes a non-renewable resource to which players have indirect access through distribution. From a technical, regulatory, administrative, political and economic point of view, the relation between access and distribution is very complex.[25] In particular, the procedures of allocation are so complicated that they have triggered a series of political and economic fights, as well as struggles for power and control among those authorities that manage access and the interest groups that need it.

c. Money

Money constitutes the basis of every move in the market. From all the goods that are to be found in economy, money is the most mobile, common and universal, as it circulates from one country to another, one continent to another, meets no borders and has no time limits.[26] It has been in particular the necessary condition for developing and applying technology, creating credibility for the alliance's name, attracting investors and overcoming market barriers.[27] The S PCS market was a market where huge investment was the necessary precondition, as the investment, obtained by big firms, has often exceeded budgets of many small countries.

**B. Strategic Actors**

Strategic actors have been important players, whose decisions and choices influence the results of economic, political and legal interactions. Such interactions have been taking place, for instance, during the international policy-making fora of the International Telecommunications Union (ITU); at meetings of regional decision-making bodies, such as European Community and the European Con-

---

309, Dahinden Erwin, Die rechtliche Aspekte des Satellitenrundfunks, Uniersitätsverlag, Freibourg/Schweiz, 1990.

[25] These procedures, which will be analysed, are initiated through engineering mechanisms, in order to calculate how much new spectrum is needed or how the existing spectrum can be relocated. Regulatory and administrative procedures, which may last up to six years, make sure that the system is going to operate without causing any interference to systems of other states. For these procedures an international regulatory body -the International Telecommunications Union- is responsible. The procedures end up with the assignment of frequencies by national governments to the systems that want to start operation.

[26] See Eric Izraelewicz, Η δικτατορία των αγορών, Ένας ιδανικός αποδιοπομπαίος τράγος, in Liousis/Ntalis, Οι διεθνείς σχέσεις στη μεταψυχροπολεμική εποχή, (The dictatorship of the markets. An ideal scapegoat, in Lioushs, Ntalis, The international relations in the post cold war era), Exandas, 1999.

[27] At this introductory point this statement may seem vague, however, it will be thoroughly analysed in the following chapters. For instance, paying for a spectrum license may cost up to hundreds of millions dollars. However it constitutes a market barrier entry, which can only be overcome by the actors that have the economic ability to pay.

ference of Posts and Telecommunications (CEPT), or through the formation, of economic coalitions for the promotion of certain technology in the market.

The term "strategic actors" has been used in a broad sense. It comprised, in particular, two large categories of actors: the ones that have been leading the economic game and have been able to develop a global strategy of networking, the "strategic economic actors", and the ones that are responsible for policy, regulation and development, the "strategic regulatory actors".

As the number of the strategic actors has increased during the past years, the study of their interaction has become more difficult. The larger the number of groups, the more difficult it has become to study co-operative or competitive patterns of interaction and monitor the strategies they develop.[28] Furthermore, due to their different nature, relations that have been market-based and have affected the emergence and expansion of global economic networks as well as regulatory structures have involved different types of organisation. These include states, firms, and regional or international organisations.

a. Economic Strategic Actors

The industrial sector has been playing a crucial role in the promotion of transborder economy. Trade and productivity have been particularly important. The choices that big industries have made, have been focusing on resource concentration, expansion and market identification for profit maximisation. Their interaction has been characterised by a dynamism expressed in various ways, as strategic actors have had their own identity and based their expansion on a network of relations, political and economic influence, as well as money.[29] Moreover, by dominating exchanges they have imposed their ideology and mentality in the market. In brief, they have chosen where to invest, where and what to produce, where to sell. In this way they have been determining the rules of the game.

Since the mid 1990s, many big industrialists have been involved in the telecommunications and S PCS game, which became a game of choices: who to team up with, which investor would be better to attract in order to obtain larger amounts of money. As a consequence, well known industialists, including Motorola, Alcatel, Loral, Hughes, Lockheed Martin, Alenia and many others, got involved in all market segments, from microchips and software to cellular mo-

---

[28] On the heterogeneous characteristics of the actors see Keohane Robert, Ostrom Elinor (ed.), Local commons and global interdependence: heterogenity and cooperation in two domains, London Sage, 1995.

[29] See Mitchel Y. Abolafia, Market as a cultural approach, in Hodgson M. Geoffrey, Screpanti Ernesto, Rethinking economics: markets, technology and economic evolution, Aldershot: Elgar, 1991, pp. 40 ff.

bile telephony and satellite communications.[30]To these industrialists it was a game of success and profit maximisation, as they could penetrate many telecommunications sectors and exploit new technologies.

b. Regulators as strategic actors

Regulators are also examined as strategic actors, as the policy they have shaped and the law they have produced has influenced the exchange of resources, the interplay among the actors and the structure of the market. Regulators have been national, like the National Regulatory Authorities (NRAs),[31] regional, such as CEPT and international, like ITU and World Trade Organisation (WTO). As the regulatory level moves from national to transnational, the number of participants increases as well as the degree of difficulty in regulating trade

There have been various methods of creation of regulatory bodies, for instance, common consensus among states and legitimisation through international agreements, (ITU), transfer of powers through legislation (European Radiocommunications Committee) or imposed through legislation from a higher institution (European Community-Member States[32]). Transfer of limited powers has been a very common way of legitimising, for instance in the case of CEPT.[33]

**C. Business: trying to succeed in order to fail - Between market and business failures**

One of the main concepts of the study is the market. In economic terms, markets are commonly defined as "self regulating exchange systems in which buying and selling takes place in a competitive bidding process between utility maximises." The market is a co-ordinating mechanism, where the forces of supply and demand are balanced.[34]The market co-ordinates economic transactions be-

---

[30] For instance, Motorola has been a manufacturer of microchips, software, cellular phones, satellite phones, heavy investor in Iridium, Hughes has been an investor in ICO and Thuraya, Loral in Globalstar, Matra Marconi in East, Alenia in Globalstar, Boeing in Ellipso, and Orbital Sciences in ECCO.

[31] The National Regulatory Authorities (NRAs) are the national, independent, regulatory bodies responsible for regulating domestic telecommunications within the European Community. The basis of the creation of the NRAs has been Article 7 of the Services Directive, (90/387/EEC, OJ No C 192/1, 24.07.90), which introduced the separation of the commercial and functional responsibilities of the National Telecommunications Organisations. For their role and competencies in the telecommunications market, see Chioni/Karageorgos, Telecom- munications in the European Union, Hmerisia, 4 January, 2003.

[32] For instance, in the case of National Regulatory Authority creation.

[33] The European Conference for Post and Telecommunication. For a thorough analysis, see Chapter IV.

[34] See Boyer and Drache (ed.), States against markets, The limits of globalisation, Introduction, p. 3. On the market mechanism, see Kit Sims Taylor, Human Society and the Global

tween actors that buy and sell, actors that are voluntarily engaged in contractual exchanges of goods and services.[35]

Markets are structures in such a way that, on the one hand, they have their own dynamic, but on the other hand, they cannot function in isolation from other social environments. This means that there exists a mutual influence between the endogenous and exogenous factors of the market. Endogenous factors are the factors or entities that are determined by the market. Representative endogenous actors are, for instance, the prices. Exogenous factors are equally important. These comprise of politics and regulation. Through these factors, the process of market evolution takes place, which leads to the process of market institutionalisation. Markets emerge, function, are successful or malfunction and fail. There has been a lot of debate on the doctrine of market failure, i.e, markets where allocation or resources are inefficient. In particular, market failure has been called the inability of a market to efficiently allocate resources so as to achieve the greatest possible consumer satisfaction.[36]

Main market failures have been:
(1) The provision of public goods. Public goods have the characteristics of non-rivalry and non-excludability. According to non-rivalry, the possession of the good by one member does not prevent its possession by others. Non-excludability means that one member of the public possessing it cannot prevent others from also possessing it.[37] When goods or services are non-rival the market leads to inefficient outcomes, because a firm has to charge a higher price than the efficient price to break even or it makes a loss and has to be subsidised. The provision of non-rival good results also in a natural monopoly, because smaller firms would produce fewer products and therefore the marginal costs per unit would be higher. Since non-rivalry just makes the market inefficient, non-excludability causes the whole market to fail. The reason for this is that a good or service is not provided when nobody pays; and nobody pays because there is no benefit from paying once the good is provided.»[38]

---

Economy, Chapter 4, at http://distanct-ed.bcc.ctc.edu/econ100/ksttext/markets/MARKETS.htm.

[35] On the various conception of the market, see Robert Boyer, The State and the Market, in Boyer/Drache, p.100.

[36] PrinciplesofMicroeconomicTheory,Chapter13, http://www.uta.edu/depken/ugrad/2306/chapt13.html.

[37] See The role of the market in the provision of public goods and services: Balancing market failure and government failure, Department of Economic and Social Affairs, Division for Public Economics and PublicAdministration, http://unpan/un.org/intradoc/groups/public/documents/un/unpan000156.pdf.

[38] Basic ideas of Microeconomic and WelfareEconomics, http://home.arcor.de/marco.wegener/essays/economics.htm.

(2) The existence of externalities for which markets are thought to be poor at compensating for.[39]Externalities can be positive (as in the case of human education) or negative (pollution). Both cases set the basis for increasing the role of the state, either for reducing negative externalities or promulgating the positive ones, as there are no efficient points.[40]The economically efficient point is where the marginal cost to all individuals in the economy is equal to market demand. But when there are externalities this social marginal cost exceeds the private marginal cost in production of the good that are borne by the firm.

(3) Imperfect information: Transactions in markets require information: identity and location of potential buyers and sellers, quality of goods or services to be exchanged and validity of property rights attached, which consumers must have. An asymmetry of knowledge, which particularly occurs where clients pay for an expertise of professionals, prevents consumers from taking a free decision. Also when consumers do not participate frequently in a market it becomes difficult to overview the price situation of a product.[41]

In the S PCS case the market mechanism of transactions and exchange of control for commodities was reaffirmed; it has not been a case of market failure. The market has been functional, but business has not been successful. In the market, information and assets have been transmitted and exchanged: firms tried to discover the preferences of the consumers so that they knew what their product should look like; they tried to predict the demand, forecast macroeconomic conditions, see what would be the moves of the competitors and at the same time advertise the product and attract a mass market. The S PCS paradigm has been a case of business failure, where the systems have showed inability to compete in the market for reasons that will later be presented and explained. In brief, the systems interested failed to predict the demands for the products and the services and the costs. As demand was overestimated, there was unnecessary expenditure on marketing and advertising strategies. Moreover, responsibility centres were decentralised, which despite being the correct course of action, were inexperienced, resulting in this way in non organised planning and non-coherent decisions. All this will be analysed thoroughly in the following chapter and in particular in the last chapter.

---

[39] An externality is said to exist "if some of the variables which affect one decision-maker's utility or profit are under the control of another decision maker". See Welfare Economics, Chapter 18, p.509.

[40] The role of market in the provision of public goods and services: balancing market failure and government failure, United Nations, New York 1999, Department of Economic and Social Affairs, Division for Public Economics and Public Administration, http://unpan1.un/org/intradoc/public/documents/un/unpan000156.pdf.

[41] Basic ideas of microeconomics and welfare economics, as above.

### 3. Satellite Communications and Satellite Personal Communication Systems: Regulating-liberalising-reregulating

Telecommunications has been a complex market, where during the past fifteen years deregulatory efforts have intensively been concentrated on two issues: the liberalisation of the market and the introduction of greater competition as well as the reduction of the power of incumbent operators through the separation of their commercial and operational powers.[42]These were attributed to independent national regulatory authorities (NRAs). The regulatory efforts started with the liberalisation of the terminal equipment sector, the provision of services (including the provision of satellite communication services, which has been liberalised since 1994)[43] and concentrated on the introduction of network competition.[44]

The legal framework that was needed to shape the function of the S PCS market had to take into consideration uncertainties and risks including the market segments which were not exactly identified, in particular vis à vis the mobile cellular market, and the untested technology.[45]As a consequence, it has been difficult to impose legally binding rules that establish concrete obligations to the interested parties relating to issues such as advanced technology regulation or proce-

---

[42] Commission Green Paper on the Development of the Common Market for telecommunications services and equipment, COM(87)290, 30.06.87.

[43] Council Directive of 24 June 1986 on the initial stage of the mutual recognition of type approval for telecommunications terminal equipment, 86/361/EEC, OJ No L 217/21, 05.08.86, Council Directive of 29 April 1991 on the approximation of the laws of the Member States concerning telecommunications terminal equipment, including the mutual recognition of their conformity, 91/263/EEC; OJ L No 128/1, 23.05.91, Council Directive 93/97/EEC of 29 October 1993 supplementing Directive 91/263 in respect of satellite earth station equipment, OJ No L 290/1, 24.11.93.

[44] Council Directive of 28 June 1990 on the establishment of the internal market for telecommunications services through the implementation of the open network provision, 90/387, OJ L 192/1, 24.07.90; Council Resolution of 22 December 1994 on the further development of the Community's satellite communications policy, especially with regard to the provision of, and access to space segment capacity, 1994, OJ C No 379, 4.

[45] Council Directive on the introduction of a pan-European land-based public radio paging in the Community, 90/544/EEC OJ L No 310/28, 9.11.90, Council Directive of 3 June 1991 on the frequency band to be designated for the coordinated introduction of digital European cordless telecommunications (DECT) into the Community, 91/287/EEC OJ L No 144/45 of 08.06.91, Council Recommendation of 3 June 1991 on the coordinated introduction of digital European cordless telecommunications (DECT) in the Community 91/288/EEC, OJ L No 144/47, 08.06.1991, Decision No 128/199/EC of the European Parliament and of the Council of 14 December 1998 on the introduction of a third-generation mobile and wireless communications system (UMTS) in the Community, L No 17/1 22.1.1999, UMTS Decision.

dural arrangements for licensing conditions and spectrum allocation.[46] As a result, European and international regulators have showed a strong preference for non-legally binding rules that comprised general guidelines and objectives.[47] Co-ordinated multilevel regulatory efforts have taken place through flexible legal schemes: regulatory efforts have taken place at all levels (supranational, national and regional) through centralised institutions (ITU, European Commissions), expert bodies (CEPT/ERC) and national authorities (NRAs-NCAs) and have had the form either of hard (Council Directives) or of soft law (ERC Decisions).

The law that emerged has been a mixture of

- generic community legislation on telecommunications, such as the one on licensing[48]
- satellite specific legislation (provision of services, frequencies) [49]
- S PCS specific legislation [50]

---

[46] These issues, which may differ from state to state, are related to space engineering and economic efficiency as well as administrative capacity. For instance, the level of economic, regulatory and administrative efficiency that the US has in performing S PCS licensing and auctions (e.g. electronic simultaneous multiple round) is very difficult to find in other regions of the world.

[47] The examples of non-legally binding rules being used by the regulators in order to overcome the obstacle of regulatory uncertainty are numerous. The more complicated the issue, the more non-legally binding the documents. For instance, see the decisions of the European Radiocommunications Committee (ERC) on the introduction of spectrum bands for S PCS and UMTS in Europe: ERC/DEC/(97)03, ERC/DEC/(97)04, ERC/DEC/(97)05 and ERC Deci sion (97)07. See also the Global Memorandum of Understanding of the International Telecom munications Union (ITU) on the free circulation of terminals for mobile satellite personal communications. These issues will be analysed in chapter IV.

[48] For instance, Council Directive of 28 June 1990 on the establishment of the internal market for telecommunications services through the implementation of the open network provision, 90/387, OJ L 192/1, 24.07.90; Directive 97/13/EC of the European Parliament and of the Council of 10 April 1997 on a common framework for general authorisations and individual licenses in the filed of telecommunications services, OJ L No 117/15, 07.05.97; Directive of 9/3/99 of the European Parliament and the Council on Radio Equipment and Telecommunications Terminal Equipment and the Mutual Recognition of their Conformity, 1999/5/EC, OJ L No 91/10, 7/4/99.

[49] For example, Council Directive 93/97/EEC of 29 October 1993 supplementing Directive 91/263 in respect of satellite earth station equipment, OJ No L 290/1, 24.11.93; Directive 98/13 of the European Parliament and of the Council of 12 February 1998 relating to telecommunications terminal equipment and satellite earth station equipment, including the mutual recognition of their conformity, OJ L No 74/01 of 12.03.98.

[50] Decision No 710/97/EC of the European Parliament and the Council of 24 March 1997 on a co-ordinate authorization approach in the filed of satellite personal-communications services in the Community, OJ L No 105/4, 23.04. 1997.

**4. Structure**

The first chapter describes the satellite technology. Presentation of technological characteristics constitute a precondition for the reader in order to better understand what have been the advantages of this new technology, why there has been such an increased industrial interest and what have been the difficulties in regulation. As a consequence, basic knowledge of technology for the further comprehension of the economic, political and legal developments is presented in this study. The concrete technology described has been considered as «new», compared to the one used twenty years ago. These technological differentiations have contributed to the formation of new balances of economical, political and regulatory forces.

The second chapter presents the S PCS market. It tries, in particular, to define it in terms of product and service provision, and to compare and analyse the degree to which it was «complementary» to the mobile digital terrestrial market. Moreover, it presents the main economic actors that have been involved in the market. These have been industrialists with local, regional and some even with global presence and power. These have been involved either as investors in the alliances formed, or as manufacturers, network operators and service providers. The involvement of a big number of actors has revealed constructions of horizontal and vertical industrial chains extending to global levels.

The third chapter focuses on the regulators, i.e., the actors that have not directly involved in the market, but have played a very important role, as they provided the legal framework that would facilitate economic activities. There has been a co-ordinated regulatory effort at various levels: international, regional and national. At international level the International Telecommunications Union has provided the forum for the provision of a global regulatory regime. In terms of regional regulation Europe is being examined within the framework of regulatory efforts that have been instigated by various institutions such as the European Commission and the CEPT assisted by various bodies and committees of expertise.

The fourth chapter analyses the regulatory framework, which governs economic activities of the S PCS market. Rules and procedures are sketched out, aiming not only at a simple presentation, but also at showing the relations between the actors presented and their emergence as products of complex procedures both at international as well as national level.

The fifth chapter analyses the strategies of the main economic players, focusing on why there has been a massive failure. It presents the strategies of the strategic alliances that have been heavily investing in the projects, as well as that of the industrialists that have been active in the sector as manufacturers, network operators or service providers individually, or have formed part of the strategic alliances. Moreover, it analyses the miscalculations that have lead business to heavy losses and failure.

# Chapter II
# Satellite Technology

## *I. Introduction*

Satellites have always played a very important role in human communications. From the first vision of a satellite orbiting the earth,[51] to the global hypernetworks, the satellite has always been a symbol of the human capacity to extend far beyond earth. During recent decades satellites have enabled numerous services to develop, including communications, broadcasting, navigation, meteorology, remote sensing and positioning.[52]The quality of these services has been increasing as the technological ability has improved, and during the last few decades there has been a tremendous evolution in satellite technology. The first satellites were passive ones and their only function was to reflect signals in space between two points on the surface of the earth. As technology advanced satellites were equipped with amplifiers and could transmit to more points of the earth. The type of satellite that followed was even stronger, enabling the signals to be directly received by many users.

The use of satellites has become particularly intense during the last few years, and has become an inseparable part of the information society.[53]Satellite systems designed for personal mobile communications, constituted a promising, emerging satellite type in the 90s for several reasons: low orbit, very fast rotation of the satellites around the earth, transmission of signals in frequency bands other than the ones already used by existing geostationary satellites, increased number of satellites for complete planetary coverage and the constant instant connection with ground infrastructure, have been the main features of this new technology.

This chapter will study the very basic notions of the satellite S PCS technology such as the satellite, earth and space segment, frequencies and orbits. It will then move on to more complicated issues, such as those of satellite constellations, intersatellite links and gateways and transmissions modes. Furthermore, it will try to give an overview of what satellite personal communications systems offer to those who what to exploit them, compared to what the existing systems do (such as geostationary satellites and mobile terrestrial cellular). This analysis is the key to understanding the chapters that follow.

---

[51] Drawings of artificial objects orbiting the earth go back to the 19th century.

[52] On the services provided, see Pelton J.N., Wireless and Satellite Communications: the technology, the market and the regulation, 1995.

[53] On information technology see Schmoch Ulrich and Thomas Schöring, Technological Strategies of telecommunications equipment and manufacturers, a patent analysis, Telecommunications Policy, 18, 5, 1994, pp. 397-413.

## *II. Satellites and their networks*

### 1. Brief historical overview

The satellite is an artificial celestial body that rotates around the earth at a certain distance and at a certain speed. The Radio Regulations (RRs)[54] of the International Telecommunications Union have defined the satellite as *«A body which revolves around another body of preponderant mass and which has a motion primarily and permanently determined by the force of attraction of that other body.»*

The satellite has been one of the means of communication, offering long distance coverage from point-to-multipoint (from one user to many users simultaneously).[55]It was not until 1957 that the first experimental satellite, "Sputnik", was launched by the Russians. The other experimental satellites that followed were launched in the early 60s and were the so-called «passive satellites».[56] These satellites were only used as reflectors in space, which meant that they did not have the ability to amplify the transmitted signals and that they could only transmit between two specific points on the surface of the earth. Since the satellites had no booster power and the signal started to weaken at 2000 miles, power ful and expensive transmitters on earth had to be installed.[57]The next and more advanced type of telecommunications satellite was the «distributor satellite», which had the capacity to amplify signals and could transmit from one point on the surface of the earth to two or more. The signals were amplified but not enough to reach the final user. They had consequently to be retransmitted from head stations to individual antennae through cable networks. More powerful sate llites made their appearance in the 60s.[58]These were the direct broadcasting sate llites, which had high transmission capacity and their signals could be directly received by the public without the intervention of an earth station.[59]

---

[54] The Radio Regulations (RRs) have been very detailed provisions concerning the use of spectrum and of orbit positions. These regulations have been elaborated by the ITU during international conferences.

[55] Theoretically, three geostationary satellites could cover the whole surface of the earth.

[56] Initially satellites where launched by the US and Soviet Union. The launch of a soviet satellite was followed by the launch of an American one and vice versa. See also, Paul Johnson, Modern Times: The world from the Twenties to the Eighties, 1983.

[57] Engelhard Helmut, Satellitendirektfernsehen in Spannungsfeld von Volkerverständigung und Propaganda, Peter Lang, Frankfurt a/M, 1992.

[58] The first direct broadcasting satellite was launched by the Americans. It was the "Early Bird" in 1965 which was placed above the Atlantic and had the transmission capacity of 240 telephone circuits or of a TV channel. For a brief historical presentation, see L.J. Libois, Genèse et croissance des télécommunications, Manon, 1983.

[59] For a detailed distinction between the three types of satellites see Ulrich Thieme, Rundfunksatelliten und internationales Recht, 1973, p.35, Andre Kerever, Copyright and space satellites Congres de la Mer Aigée, 1983, p.148, Fawcett, Law of Outer Space, 1984, Adolf

**2. The satellite network**

The satellite can be described as a space transmitter, or better, a type of mirror placed in space which reflects and strengthens the transmitting signals, while being placed in a certain orbit position and at a certain frequency.

The satellite forms part of the satellite system, which consists in very general terms, of the space segment and the earth (ground) segment. The term "space segment" refers to the satellite and to all the other equipment that is placed in space and that is necessary for the operation of the satellite, such as the satellite's tracking, telemetry, command, control and monitoring systems. It is the central unit of the satellite system. The "earth segment" comprises of the installation on the surface of the earth, including any additional equipment. It forms the interface between networks and users.[60]

***III. Frequencies-Orbit***

**1. Frequencies**

Spectrum is a "scarce" resource and has two technical specificities.[61]On the one hand, spectrum is an indefinite, renewable resource. On the other hand, it is divided into portions, each with a specific use. This means that once a spectrum portion is occupied by one user, it cannot be occupied by another.

---

Dietz, Satellite Communications and Copyright Law, pp.81 ff., in Telecommunications Satellites and International Law, Colloquium, 1989.

[60] The satellite receives signals transmitted by an earth station at a certain frequency. After reception, the signals are transmitted by the satellite with the help of a transponder back to earth. The systems of telemetry, control and monitoring ensure the control of the satellite vis à vis other satellites. The term transponder is used for the mechanical part of the satellite on the platform, a sort of antenna, through which the transmission takes place. It is very important that during the transmission the satellite keeps a certain position. First, so that the interested states receive the transmission and second, so as to avoid the possibility of collision with other satellites. See, on the technical capabilities of satellites, Tony Reed, Where will satellites win in the 90s?, in Schorrock, (ed.) European Communications, Technologies and regulations of the single market, Blenheim on-line publications, 1989, p. 65 ff. Besides the categorisation to earth and space segment, the ITU and the national governments divide satellite systems into three distinct parts: the uplink, the satellite, and the downlink. The uplink refers to the transmission of a signal by "channelling" them towards the satellite, whereas the downlink describes the channelling of the earth signal as it descends from the satellite to the receiving earth station.

[61] Spectrum has also been characterized as a "limited" resource. However, this characterisation does not completely correspond to physical reality, as spectrum is not limited, but hard to access by many simultaneous users. Therefore "scarce" is more accurate. Also Christian A. Herter, The electromagnetic Spectrum: A critical natural resource, in Natural Resources Journal, 1985, pp. 652-663.

The difference between the various forms of electromagnetic energy is their frequency.[62] A frequency is a measure of the number of radiated waves passing a given/fixed point in a certain unit of time.[63] The same frequency can be used by the same users at the same time but in a different area, or in the same area but at a different time. The same frequency at the same time and in the same area can not be used.[64] As spectrum is needed in all telecommunications, the demand is so high that the existing spectrum does not appear to be large enough. Furthermore, the smaller the spectrum, the bigger the problem of interference among different services using the same or neighbouring bands.

The higher levels of electromagnetic energy are visible lights, X-rays and gamma rays whilst the lower levels are those of radio waves. For the satellite transmissions, the most important frequency bands are those between 1 and 30 Ghz. These are the L-Band (1-2Ghz), S-Band (2-3Ghz), C-Band (7-8Ghz, where telecommunications satellites transmit), X-Band (7-8 Ghz), Ku-Band (10-19 Ghz, in which broadcasting satellites transmit) and finally, the Ka-Band (17-31

---

[62] The signals transmitted by satellites are electromagnetic. Electromagnetic signals are invisible, they travel at the speed of light and are capable of travelling enormous distances. The complete range of rates at which electromagnetic waves fluctuate as they pass through space is comprised by the spectrum. In the spectrum the values of the electromagnetic energy differ.

[63] The unit of measurement for frequencies is the "hertz". The concept of the hertz is that of a cycle. When a wave passes completely through a given point it has completed a cycle. Consequently, in order to measure the waves that pass through a certain point it is enough to measure the number of cycles that pass a given point in a second. A kilohertz is a thousand cycles per second whilst a megahertz is a million cycles per second. The range of frequencies used by broadcasting organisations, radio stations or ground stations can be considered to be the channel of the transmission. Various categories of frequency bands have been elaborated: the longest (myriametric waves) waves (3-30Khz), long (kilometric) waves (30-300Khz), middle (hectometric) waves (300-3000Khz), short (decametric) waves (3-30 Mhz), ultra short (metric) waves (30-300 Mhz), decimetric waves (300-3000 Mhz) centimetric waves (3-30 Ghz), millimetric waves (30-300 Ghz), decimillimetric waves (300-3000 Ghz). The most important band for this study is that of decimetric waves, since the new satellite personal communication systems operate in this band. Schertler Ronald J., Gedney Richard T., Zernic Michael J., ACTS' legacy: The Ka-Band explosion, Aerospace America, February 1996, p. 33, Herter Christian, The electromagnetic Spectrum, Natural Resources Journal, 1985, p.25, p.652-663, Horst Kuchling, Taschnenbuch der Physik, 16th Edition, 1996, European Commission, The impact of the new technologies on the industrial countries, Final Report, 1988, p.40ff., Delcros Bertrand/Trouchet Didier, Controverse, les ondes appartiennent-elles au domain publique? R.F.D.A., 5, 2, mars-avril, 1989, 251-258.

[64] Time, area and frequency bands form the TAS package, as deVany et al., have called it in de Vany et al. A property system for the electromagnetic spectrum Stanford Law Review, v.21, n.6, p.1499 ff., 1985.

Ghz).[65] Filings have been made by satellite systems for allocation in the V-Band (45-60 Mhz).[66]

**2. The orbit**

A satellite can only transmit if it has a certain orbit position. The RRs of ITU describe the orbit as

".the path relative to a specific frame of reference, described by a centre of a mass of a satellite or other object in space subjected primarily to natural forces, mainly to the force of gravity".[67]

The orbit in which the majority of the satellites have - up to now - been rotating the earth is the so-called geostationary orbit. In the geostationary orbit satellites rotate around the earth at a height of 36000km above Equator.[68]The rotation is completed in 23 hours and 56 seconds. If seen from a point on the surface of the earth, the satellite always seems to be stationed at the same fixed point on the horizon.[69]However, new technology has allowed satellites to be operational in other orbits as well. These are the low earth orbit (LEO), placed 800 Km above the surface of the earth, the medium earth orbit (MEO) at a height of 100.00Km and the elliptical (EEO) and high elliptical orbit (HEO) at heights of 400 and 30.000Km respectively.[70] The satellite constellations, which this study is going to examine, have been rotating mainly in the Low Earth Orbit.[71]

---

[65] COM (90) 490 Satellite Green Paper, p.27.

[66] Communications Systems limited and Hogan & Hartson, Prospects for personal satellite communications using service links in the Ka-Band, Report for the European Commission, August 1997.

[67] Marika Taishoff describes orbit as a "doughnut shaped volume of space centered around the earth's equator", in State responsibility and the direct broadcasting satellites, 1987.

[68] See Stephen Gorove, The geostationary orbit, issues of law and policy, 73 American Journal of International Law, 1979, 444-463; Gehring, Geostationary Orbit, Technology and Law, in proceedings of the 19th Colloquium on the Law of Outer Space, 173, 1978.

[69] The term "geostationary" is often confused with the term "geosynchronous". Geosynchronous means that the orbit of the satellite is equal to the sidereal rotation of the earth, whereas geostationary means that the orbit (which lies in the equatorial plane) and the rotation of the satellite have the same direction as that of the earth.

[70]There is also the Near Earth Orbit, at a height of 150-450 Km above the surface of the earth. Because of the influence of the atmosphere this orbit results in a very short satellite lifetime and is therefore avoided by the satellite systems. See Mosteshar Sa'id, Development of the regime for the Low Earth Orbit and the Geostationary Orbit, in Lafferanderie, Outlook on space law over the next 30 years, p. 81ff.

[71] See a very analytical presentation of satellite technològy and function in Bruce R. Elbert, Introduction to Satellite Communication, Artech house, 1987.

## *IV. Satellite personal communication systems: the advanced technology*

### 1 The concept of personal communications

Wireless communications have been evolving towards broadbandness and integration.[72]This process has started with voice telephony (fixed telephone lines) to include all services (f.e, fax, data processing, e-mail, videoconferencing). The concept of personal communications was originally developed by the mobile industry to facilitate personal mobility as well as terminal mobility and allow people to use "phones on the move".[73]The main point of this concept has been the ability to access and be located while in motion, without being hindered by any kinds of obstacles. This has required some form of wireless user-interface in the network. In particular, PCS (personal communications systems) have been considered by the mobile industry to be an umbrella for wireless network access, as they should offer mobility through a variety of different technologies.[74]In general terms, a PCS network is a high capacity system, which provides two-way services to users who do not have a fixed geographical location. It is characterised by a high degree of mobility for users and/or terminals.[75]Moreover, personal communications are communications where the customer has a single number, a single supplier and, where possible, a single bill.[76]Furthermore, personal communications are signalling the convergence of fixed and mobile services.

There have been a number of different definitions of PCS and these definitions can be divided into two groups. One group has defined PCS as an overall concept of personal telecommunications for all kinds of networks, the other group has restricted PCS to mobile networks only. However, PCS are not only mobile, but also fixed; and as a consequence, the broad definition of the ETO (European

---

[72] Broadband means higher frequencies, which means that the higher the frequencies, the larger the amount of data that can be transmitted. Integration means the technological tendency towards fixed-mobile integration.

[73] ITU definition: Personal mobility is the ability of the user to access telecommunications services at any terminal on the basis of a personal identifier, and the capability of the network to provide those services delineated in the user's service profile. Personal mobility involves the network capability to locate the terminal associated with the user for the purpose of addressing, routing and charging (ITU-T Rec. F.851).

[74] See European Commission, Satellite Communications and their consequences for European telecommunications, trade and industry, Main Report, March 1994, p.19.

[75] This distinction made between the terminal and the user is necessary in order to allow mobility of the user across several networks. By using the Subscriber Identity Module (SIM), mobility can be achieved in order to access personal services from any terminal, either from a cellular or from a satellite system.

[76] Technological convergence is expected to comprise of the following levels: network integration (network intelligence in order to identify the current location of the user, the preferred method of delivery and the billing, transport infrastructure and access, which is dependent on radio spectrum for mobile and on copper or fiber for fixed networks.) and terminal convergence.

Telecommunications Office) best describes the nature of these services.[77] "Personal Communications Services" has been used as a generic term for services which provide person-to-person calling, independent of location, terminal use, the means of transmission (wired or wireless) and/or the choice of technology.[78]

## 2. The concept of Satellite Personal Communications

Satellite personal communications (S PCS) constitute one of the networks of personal communication systems. They are, in particular, the satellite component of personal communication systems that allow communications everywhere and at anytime, and provide personal communication services to users wherever they are, static or on the move. They are systems, which provide universal or regional communication services, a function that includes terminal mobility provided by wireless access,[79] personal mobility based on personal numbers,[80] and service portability through management of user service profiles[81]

In the relevant decision on "Iridium", the European Commission gave a broad definition of the S PCS project and consequently of LEO systems, by describing them as systems that:
"..provide global digital wireless communication services using constellation of LEO satellites. Services will include voice telephony, paging and basic data services (such as facsimile) and will be provided via portable hand-held (dual mode

---

[77] ETO Final Report on numbering in S PCS, November 1997.

[78] The Federal Communications Commission has also given a very broad definition. It has defined PCS as a family of mobile or portable radio communication services with the ability to provide services to individuals and businesses. This definition is broad enough, in order to comprise not only mobile, -since many define PCS as mobile, which is not true-, but also portable, which means that they may be also fixed. See Warner Lisa M., Wireless technologies, creating competition in the local exchange market, in Common Law Conspectus, Winter 96, vol. 4, pages 51-78. Also Thomas A. Monheim, Personal Communications Services: the Wireless Future of Telecommunications, in Federal Communications Law Journal, 1992, vol. 44, n. 2, p. 336 ff. PCN services are services based on intelligent networks, that is cordless radio-telephonic networks based on digital and microcell technologies. On this issue see also: Steven Titch, Is PCS more than just a portable pocket phone? Telephony, June 17, 1991.

[79] This function refers to the ability of a mobile terminal to access telecom services from any location and to be in motion. The network is able to locate and identify the mobile terminal on the move.

[80] Each user has a unique personal identity (a personal number), and on the basis of this number the network can identify the user as he moves. In this way the users may receive and make calls and access subscriber telecommunications services on any terminal in any location , see John R. Wickens, Who will be the winners? The making of a top player, Pan Asian PCA 96, January 1996.

[81] This capacity refers to the network capacity to provide subscribed services. See Raj Pandya, Emerging Mobile and Personal Communication Systems, IEEE Communications Magazine, June 1995.

or single mode) telephones, vehicle mountained telephones and other subscriber equipment."[82]

### 3. The technology

As technology has made possible the exploitation of orbits other than the traditional GEO one, the satellite PCS have mainly been low (LEO) and medium earth orbit (MEO) systems.[83]Non geostationary orbit satellite systems (NGOSS) have been proposed as a means of providing high-quality, low-delay communications to handheld portable devices, usually in remote areas with no ground communications infrastructure, anywhere in the world.

### A. LEO Systems

Many of the S PCS proposed have been LEO systems, the analysis focuses on mainly LEO satellites. LEO satellites are distinguished in three categories, related to the frequencies they transmit and the services they offer: the Big LEOs, Little LEOs and Giga LEOs. These three types differ mainly in the size of the satellites used, the frequencies they transmit, and their applications. Big LEOs are defined as "non-geostationary low earth orbit systems, providing global narrowband voice telephony".[84]They have been assigned the following frequency bands: 1610-01626.5 Mhz, 2483.5-2500 MHz, 1980-2010, 2170-2200 MHz and their function has been to extend services from terrestrial fixed to mobile systems, providing mainly voice and data services. Little LEOs have been "low-cost, low-bitrate real time store and forward messaging systems suitable for tracking, monitoring applications and wireless e-mail"[85].They use frequencies in the highly congested bands of below 1 Ghz. Mega LEOs (or Giga LEOs) are broadband multimedia satellites of high capacity that promise to offer high speed interment and on-line services and infrastructure links for telephony and mobile communications using frequencies 20/30 Ghz.

**Table 3. Schematical presentation of the three types of LEO systems**

| *System Type* | *Little LEO* | *Big LEO* | *Giga LEO* |
|---|---|---|---|
| *Example* | ***Orbcomm, Vita*** | ***Iridium, Globalstar, ICO*** | ***Teledesic*** |

[82] See Decision of the Commission of 18.12.1996-IV/35.518-Iridium, ABIEG Nr. L.16 v.18.1.1997, s.91 Tz.25, also König/Zeiss, EG-Telekommunikationsrecht und Lizenzierung von satellitengestützten persönlichen Kommunikationssystemen, EuZW, Heft 5/1999, p. 133 ff.

[83] Also called "non-geostationary orbit satellite systems"(NGOSS). Elliptical and high elliptical orbits have not yet been exploited, as technology is not yet complete for such applications.

[84] See European Commission, Satellite Action Plan, Activities Report, SEC (99) 290, p. 7.

[85] Satellite Action Plan, Activity Report, SEC (99)290, p. 8.

| *Terrestrial Counterpart* | ***Paging*** | ***Cellular*** | ***Fiber*** |
|---|---|---|---|
| *Frequency* | *<1 GHz* | *1-3 GHz* | *30/20 GHz* |

Giga LEOs are not considered to constitute S PCS *stricto senso*[86], since S PCS have only been used for voice and data transmissions, which means that they only included Big LEOs and Little LEOs.[87]Moreover, Giga LEOs, transmit in frequencies above 3 Ghz and have not been, at least from a de jure point of view, considered to be S PCS.

**B. LEOs vs. GEOs**

On the one hand, these two types of systems hold common characteristics. Satellites can cover the whole of the earth's surface and are invulnerable to surface calamities, such as earthquakes, floods, fires and hurricanes. Furthermore, as the distances between users and the variability of the traffic increase, satellites' wider area of coverage can make them cost-effective. On the other hand however, they have considerable differences.

One of these differences is the orbit. Geostationary satellites move in the geostationary orbit, whereas LEOs and MEOs move in orbits that are lower and closer to the earth. Rotation in the geostationary orbit, compared to rotation in other orbits has a number of advantages. Geostationary satellites always appear to be at the same point on the horizon, whereas in other orbits, because of the satellites' high speed, earth stations with traceable antennas are needed to track satellites across the sky. Moreover, for communication to and from one particular point on the earth, more than one non-geostationary satellite is needed. Once more, this is caused by the high speed at which non-geostationary satellites rotate around the earth. The shorter the rotation time and the higher the speed, the greater the number of satellites needed becomes, since one must always remain above the same point of the earth's surface, so that communications above this point can be provided.

---

86 *Stricto Sensu* in the context of the S PCS Decision, as analysed below in Chapter IV. The decision considers S PCS the system that transmits voice and data and uses frequencies up to 3 Ghz. See p.11 of the Decision 710/97, OJ L No 105, 7.5.97.

87 And MEOs, as long as they use frequencies 1-3 Ghz or even below 1 Ghz (Little LEOs). This "definition" is contained in the S PCS decision. Various types of categorization of satellite personal communication systems have been suggested, like their orbit, the weight of the satellite and their application. See Goueff, Stephan le, Licensing global mobile personal communications by satellite: the quest for Holly Grail? Annals of Air and Space Law, 1997, 22, 1, p. 417-431. Taking into consideration the Decision of the Council 710/97, it is however suggested that the most important issue is the frequencies used for the transmissions.

Furthermore, in the geostationary orbit, satellites need a lot of power to send the signal a distance of 36000 Km and therefore large terminals are required for its reception. On the contrary, LEO systems require small reception terminals, since the distance the signals have to cover is much shorter. In addition, another consequence of using different orbits is latency or signal delay. Non geostationary systems, because of their low orbit and the speed of their rotation were seen by industrialists as the best way to avoid latency and echoes. These are natural phenomena that interfere with phone calls and cause delays of up to half a second in telephone transmissions (as the signal transmitted from a geostationary satellite travels 22.320 miles to reach the satellite and then to go back to earth.)[88]

There are other technological issues that also differentiate the non geostationary satellite systems NGSOSS from the already existing geostationary ones, such as payloads, channel characterisation, space segment, gateway earth stations, switch and network and the whole system costs.[89]However, in general terms, as regards costs, (from the point of view of the satellite technology) NGOSS are much more costly, whereas the GEO satellites are costly from the point of view of transmission's and the technological equipment of reception.[90]

### *V. The S PCS Network*

#### 1. Constellation-Intersatellite Links

LEOs and MEOs are based on satellite constellations. A "constellation" describes a number of satellites placed at different or adjacent planes and at certain positions, so that at least one satellite can be seen from every point on the earth. Each satellite serves as a processor, demodulating and decoding all received data packets, storing them in digital memory, and then encoding, modulating and re-

---

[88] In particular, as far as delays in round-trip communications are concerned, an echo has a minimum of a 500 m/sec delay, whereas LEOs have less than 30 m/sec in round-trip communications. Abbas Jamalipour, p.12.

[89] The payload is the satellite's sub-system, which accounts for 30% of the satellites dry mass. Their components are antennas and transmitters, filters and multiplexers, using amplifiers at L and S bands. The channel characterization is necessary in order to design the S PCS system: measurements, elevations, narrowband or wideband requirements. Modulation and multiple access schemes are affected by various factors, such as the constellation altitude, frequency band, service quality, system capacity. The system cost is the sum of the space segment, the gateway earth stations, the subscriber unit, switching and networking. The space segment cost is determined by the total radio frequency power transmitted. Switching and networking determine the complexity of the network control: the more the LEOs or the MEOs, the more gateways are needed in order to control the satellite system.

[90] See Klodt Henning, Wettbewerb und Regulierung in der Telekommunikation, Tübingen, Mohr 1995.

transmitting the signals.[91] The number of satellites that form the constellation may vary, from six to several hundred. The main features of these systems, besides the narrow bandwidth, are the very fast response times and the global connectivity as well as the complexity of the satellite hardware and of the network control.

There have been S PCS systems which have used intersatellite links (ISLs).[92] This is a new element of satellite technology, which has made constellations form self-contained networks in space. ISLs may reduce the costs for terrestrial public telephony networks and make a satellite system completely independent by increasing the reliability of satellite transmissions in cases of natural disaster on the earth. However, intersatellite links have not been used by the majority of the non geostationary constellations, because of the technological complexity and their cost: from a technological point of view it has been difficult because of the large distances between the satellites in space and the changing of their positions, which require antenna steering.[93]

**2. Gateways**

Another element of the S PCS has been the gateways. The gateways, which form part of the system infrastructure, are ground stations which act as packet relays and regional control earth stations and allow the interconnection of the satellites with the terrestrial network that is the public switched telecommunications network[94], which make the connection possible between the system and other normal telephones. Gateways act as the interface between the satellite constellation network and the terrestrial fixed network. They are also likely to act as the control signals of the satellites for attitude monitoring, station keeping, and internal housekeeping functions. A gateway «sees» one or more satellites in the

---

[91] On S PCS technology, see Lloyd Wood, Network Performance of non geostationary constellations equipped with intersatellite links, Msc Thesis, November 1995, to be found in http://www.ee.surrey.ac.uk/Personal/LWood/Publications.

[92] For instance, Iridium and ECCO. There are in particular two types of intersatellite links. The first one is the link between satellites on the same orbital place, called intraorbit intersatellite links. The second type is a link between satellites on different orbital planes, called interorbit intersatellite links.

[93] See Abbas Jamalipour, p. 45, also Comsys, Hogan & Hartson, prospects for personal satellite communications, using service links in the Ka-Band, August 1997.

[94] Public Switched Telecommunications Networks (PSTN) form part of the public telecommunications infrastructure, which permits the conveyance of signals between defined network termination points by wire, microwave, optical means or by other electromagnetic means. Contrary to private networks, they are open to all users who want to be connected to the network. They include terrestrial wire, microwave, mobile and satellite networks. See Oliver Stehmann, Network Competition for European Telecommunications, Diss, EUI, Florence, 1993.

constellation at all times in order to ensure that it can pass connections between the terrestrial and space networks. [95]

**3. Mobile Control Centres/User Terminals**

Mobile control centres serve as earth stations for tracking, telemetry and control of the satellite constellation, as well as for the communications network. They also communicate both with the satellites and the gateways. The only other piece of equipment of the network infrastructure left to mention is the handheld terminal for the end users. These terminals are pocket sized and they can be carried and used everywhere. Moreover, the new S PCS have made the connection of these terminals to other terminals of land-based wireless telecommunications systems possible, provided that new roaming conditions for interconnection between the systems are elaborated.[96]The market has incorporated ground cellular network capability, so that the satellite constellation network is only used when the ground network is unavailable.

**4. System operation**

In "traditional" (geostationary) satellite transmissions, a ground station on earth instigates the message and transmits it (uplink) to the satellite. The satellite receives the message, converts it to another frequency and transmits it back to earth (downlink). In this way re-transmitted messages do not interfere with incoming ones.

In the S PCS, transmission takes place through a more complicated technology. The Iridium system, for instance, was co-ordinated by 12 physical gateways distributed around the world. Intersatellite links operated in Ka-Band.[97]The user dialled a phone number with the handset using an international 13 digit number, as one normally would do using a standard telephone. The user pressed the send button to access the nearest satellites. The system identified the user's position and authenticated the handset at the nearest gateway with the home location register. Once the user was validated, the call was sent to the satellite. The call was routed through the constellation and dropped to the gateway closest to the desti-

---

[95] A number of gateways, spread world-wide, are likely to be necessary to handle all of the inter-network load and to keep connections. Theoretically, a constellation network could be fully functional with only one gateway, provided that the gateway and ISLs have sufficient capacity. In fact the existence of ISLs makes the existence of more than one gateway unnecessary, whereas systems with no ISLs have to use up to 80-100 gateways.

[96] Roaming agreements are of pure technical nature and deal with the elaboration of technical standards for interconnection between the networks.

[97] From 23.18 to 23.38 Ghz and satellite gateways links operate in Ka-Band at 29.1 to 29.3 Ghz (uplink) and 19.4 to 19.6 Ghz (downlink).

nation. There it was completed over standard terrestrial circuits.[98]For a call from a fixed position the process was reversed. In this case, the call was placed and as the system identified the receiver, the handset rang wherever the user was on earth. This meant that there was no need for the satellite to "see" gateway and user simultaneously, whereas this is a condition for the function of Globalstar.

Globalstar, the second Big Leo operational system, provided low-cost, high quality telephony and other digital telecommunication services such as data transmission, facsimile and position location to areas currently underserved or not served by existing wireline and cellular telecommunication systems. The user was able to choose between single and dual mode terminals allowing fully automatic switching between satellite and terrestrial mobile personal systems to provide the best and most economic service.[99]

Globalstar has been a "bent pipe" system, meaning that the strong ground infrastructure substituted the missing intersatellite links. The call followed the path mobile/satellite/ground stations/terrestrial network/ground station/satellite /mobile. The existence of many ground stations has been necessary so that each satellite could see a ground station in its footprint, and therefore passed calls to the end users.[100] The cost and construction of these stations would offer the advantage of cheap satellites to untested technology.[101]

---

[98] See Robert A. Nelson, from Concept to reality, Via Satellite, September 1998.

[99] See Rob Frieden, Satellites in the Wireless Revolution: The need for realistic perspectives, Telecommunications, June 1994.

[100] In the satellite systems, the satellite is the dominant feature, whereas in the cellular systems, the areas to be served are divided into "cells", each with a base station and radio tower. Cell phones transmit to the nearest base station on one frequency and receive from a different frequency. While making a call, the subscriber passes from one cell to another. See Evans, John, New Satellites for personal communications, Scientific American, April 1998.

[101] On the technical issues of S PCS and satellite network constellations see Lloyd. M, Ananasso F., Vatalaro F., (ed.) Mobile and Personal Satellite Communications, Springer Verlag, 1995, 174-201, Egami S., Satellite link requirements in personal satellite communications, Space Communications, no. 12, 1994, pp. 105-114, Lawrence M. H., Patterson D. P., Stuart J.R., Tuck, E. F., The Calling Network: a global wireless communications system, International Journal of satellite Communications, vol.12, 1994, pp. 45-61, Leopold R.J., Low Earth Orbit global cellular communications network, Journal of Astronautical Sciences, Vol. 33, April-June 1985, p. 147-161, Walker J. G., Some circular orbit patterns providing continuous whole earth coverage, Journal of the British Interplay Society, vol. 24, pp. 369-384, 1971, Pickholtz R.L., Communications by means of low earth orbiting satellites, Modern Radio Science 1996, Miller, M.J., B. Vucetic and L. Berry, (eds.), Satellite Communications: Mobile and fixed services, Boston: Kluwer Academic Publishers, 1993. Also Gardiner J.G., West D., Personal Communication systems and technologies, Artech House, 1998, Fudge R., Gardiner J.G., The future of third generation mobile systems, Artech House, 1995, Barton S.K., Introduction to CDMA, CNTE publication, August 1993.

# Chapter III
# The Satellite Personal Communications Market

## *I. Introduction*
### General Remarks
The construction and operation of satellite systems that could provide mobile services through the use of particular technology (LEO satellite constellations, intersatellite links, gateways, handheld terminals) and that could make the provision of high speed voice and data services available throughout the globe, was believed to be beyond real expectation.

The first S PCS that became operational, offering services to customers, was Iridium in 1999, followed by Orbcomm and Globalstar. According to the industrialists, the demand for satellite personal communication products would increase and the market appeared at the beginning to be very promising. As a consequence, their interest started to grow for two reasons. First, the new satellite technology would reduce the high ground infrastructure costs that the GEO systems required. Second, the provision of mobile satellite voice and data services would become possible: they would be able to compete with those of the mobile terrestrial cellular systems and in addition they would have the advantage of global coverage. In this way profit was guaranteed.[1]

This chapter will focus on the S PCS market, its nature and structure on the basis of technological differentiation with regared to product, as well as the services provided - satellite constellations rotating in lower than the geostationary earth orbits and offering a variety of voice and data services. Moreover, the economic actors involved will also be presented: players that have been active at the level of investors, manufacturers, gateway operators and service providers. Interesting interactions have been taking place in particular at investors' level, where industrialists have been forming strategic alliances and investing huge amounts of money in the systems. This interaction has not only been taking place at regional level, but also at global and as a consequence, worldwide horizontal and vertical chains of co-operation have been formed. In addition, systems that have been active as well as systems that were or are expected to play an important role, will be examined: Big Leos (Iridium, Globalstar, ICO), Little Leos (Orbcomm), Regional S PCS (Thuraya, East) and Giga LEOs (Teledesic).

## *II. The satellite market*
### 1. Historical Background
The satellite market emerged at the end of 1960s and beginning of 1970s. Initially, satellites were launched by the US and Russia, on an experimental basis, for space observation, military and voice applications. Due to their limited capacity,

the satellites were initially only «passive» reflectors of electromagnetic signals in space. However, the more the technology improved, the more their value as a communications means was appreciated. As a consequence, in the 1970s and 1980s, satellites have been widely used for broadcasting, advanced fixed (FSS) and mobile applications (MSS). In particular, the FSS (Fixed Satellite Services) market was the first market to be developed in the early 1960s.[102]This was the so-called "traditional market", where the satellites transmitted between fixed points on the surface of the earth. They were primarily transmissions of voice and data. In the 1970s, satellites started being used for broadcasting applications. The direct-to-home and direct broadcasting satellites constituted the biggest market sector where satellite technology has been used, as 80% of the satellite transponder capacity has been used for broadcasting purposes. The providers of such services were INTELSAT, EUTELSAT, SES Astra, Hispasat.[103]

In the 1980s there were important developments in the satellite technology and services. Besides broadcasting applications, useful types of FSS for business communications emerged. These were systems such as SBS (Satellite Business Systems), IBS (International Business Satellite) and SMS (System of Satellite Multiservices), VSATs (Very Small Aperture Terminals).[104]Moreover, mobile services emerged mainly provided by INMARSAT,[105]enabling in this way

---

[102] Matra Marconi, Communications Satellite Systems beyond the year 2000, Report prepared for the European Commission, 1991.

[103] In Direct-to-Home (DTH) television services Europe has been the world leader and growth is still expected.

[104] In general terms, these are systems of transmitting information from a fixed system to a satellite and then to another central reception antenna, which then distributes to other receivers. See Georgia Chioni, Regulating Frequencies, Actors, policy-making and legislative framework in the field of mobile and satellite communications, Aristoteles Law Review Series, vol. V, Esperia publications, 1999.

[105] INMARSAT (International Maritime Satellite Organisation) has been one of the three international satellite organisations, the other two being INTELSAT (International Telecommunications Satellite Organisation) and EUTELSAT (European Communications Satellite Organisation). The International Satellite Organisations have operated as commercial entities, however, on an intergovernmental basis. INMARSAT provides space segment capacity for mobile satellite services, maritime TV transmissions, voice, data, fax and aeronautical services. Analytical on INMARSAT, EUTELSAT and INTELSAT their past regime and the reform, see Chioni G., Globalsiation of Personal Communications, The Satellite Personal Communications Case- Regulating a global interplay of actors for a market or a failure? PhD, EUI, Florence, 2000, Satellite News, vol. 21, no.39, 28 September 1998, p.9. Public Network Europe, ISOs: spun out of orbit, June 1999, p.28, ISOs: spun out of order, PNE, June 1999, p. 32, Johannes P. Pfeifenberger, Hendrik S. Houthakker, Competition to international satellite communications services, in Information Economics and Policy, 10, 1998, 403-430.

communication between "points on the move", such as ships, aeroplanes, space stations.[106]

During the 90s with the improvement of technology, satellite applications have multiplied in particular in the telecommunications and the broadcasting market, as well as in military and space defence industry.

**2. The nature of the European satellite market: main features**

Satellite communication systems evolved at a rapid pace during the previous decades. However, the satellite market has only been a marginal part of telecommunications, representing only less than 1% of telecommunications market,[107] with the sole applications being one-way data and TV distribution. There have been various reasons for this. First, during the past decades, the satellite market has faced severe competition from the terrestrial networks with the introduction of the fiber-optic networks in long distance communications (transcontinental). In fact, the fiber optic became the key technology for major investors in the international telecom market, such as MCI and US Sprint. In the second half of the eighties, for example, investment in fiber optic reached than $15 billion, whereas in the satellite sector it reached no more than $1 billion.[108]

Second, the transmissions received by each of the Member States, comprised in the satellite's coverage area, have not always been welcome, due to cultural, political, linguistic, economic, or administrative differences.[109]As the footnote of

---

[106] See Telecommunications for Europe, the CEC resources, 1992, p. 230 Mobile Communications have been both terrestrial and satellite. Mobile land-based cellular communications comprise services such as the ones of cellular radio, DECT, GSM, ERMES, pagers, whereas mobile satellite services have up to now been an exclusivity of Inmarsat. See Pelton J.N. Wireless and satellite telecommunications: the technology, the market and the regulation 1995.

[107] Analysis, The impact of non-European competitors on a deregulated European Satellite telecommunications market, Report for the European Commission, January 1993, p.1.

[108] Economies of satellite communications, Telecommunications Policy, 1994, 18, 6. The optical fiber has been the attractive solution to investors seeking lower costs, as it offered a far greater capacity for a given thickness of cable and better potential for expansion. The installation costs have been much less than the ones for coaxial cables. Furthermore, optical fiber cables have been very reliable, as they have been immune to electromagnetic interference and crosstalk. These advantages have made fibers a serious competitor to telecommunications satellites. See Annemieke J.M. Roobeek, Telecommunications: An industry in transitions, in H.W. de Jong (ed) the structure of European industry, 2nd, 1988, Kluver Academy Publishers, p.309.

[109] A distinction has to be made between the service and the coverage area. Service area is called the area to which the transmissions are deemed to be received. Coverage area is the area, in which the transmissions are received. It may be much bigger than the service area be-

the satellite is big, Even if the service area includes two or three countries, the coverage area may comprise of a number of neighbouring states, which have diverse societal and cultural conditions.

Third, the satellite market has not been homogenous, as there have been varying degrees of economic development of telecommunications infrastructure. In addition, because of the geographical overlapping of the various satellite systems and the shift of jurisdiction from the domestic to EC level, various political and regulatory problem were raised, which consequently made the satellite market unattractive to important players.

Due to the constant evolution of the satellite technology (worldwide coverage, independence from weather conditions and geographical anomalies) and to the increased interest of powerful industrialists, during the past years the sector has been constantly expanding. In particular, in recent years, the satellite communications industry has been driven by very strong presences: not only the traditional ones such as Hughes, Space Systems Loral, Matra Marconi, Mitsubishi, Toshiba, and Lockheed Martin but many newcomers as well. These have been very ambitious entrants, such as Motorola, Boeing, Raytheon, TRW, Alcatel, NEC and new powerful strategic alliances.[110]

The satellite market has been expected to increase from up to 10% of the total telecom market within the coming years. Applications have been multiplied. The increasing demand of customers for entertainment programming (TV), mobile communications and access to high bandwidth internet data have created new opportunities for satellites. This is why new technology has been installed at a rapid rate. This new technology has included board processing and switching, more efficient solar cells, higher power components, more efficient heat dissipation techniques and large antennas. The emergence of the S PCS and the under evolution of the broadband satellite market have revealed a new dimension of the dynamism of satellite communications.[111]

---

cause of the footprint of the satellite. Also, Euroconsult, Economic impact of competing space segment, Final Report and Summary, March 1993.

[110] For instance, Boeing/Motorola/Teledesic, Hughes/Panamsat. See Josef N. Pelton and The Global Satellite Communications Review Panel, Global Satellite Communications Review, Via Satellite, September 1998p. Players in the satellite market have had different backgrounds, being: i. telecom operators, some owing their own satellites, some with a purely domestic focus, some operating away from their home basis. (f.e. France, Germany, Italy), ii. terrestrial value-added network operators (Reuters, GEIs, EDS), iii. fully integrated space equipment manufacturers (Hughes), partially integrated players (Alpha Lyracom/Panamsat), iv., international treaty organizations (Inmarsat, Eutelsat). See Analysis, Report on Satellites, 1993.

[111] Ibid, p. 33.

### *III. Defining the S PCS market*

#### 1. Relevant Market

The rapid technological evolution has had a profound impact in the telecommunications market. As a consequence various sub markets have been created and many sectors identified, among them:[112]

- international voice telephony services,[113]
- advanced telecommunications services to corporate users,[114]
- standardised low-level- packet-switched data communications services and resale of international transmission capacity,[115]
- global communications services, internet- access services to end users,[116]
- Universal service connectivity,[117]
- seamless pan-European ,mobile telecommunications services to internationally mobile customers,[118]
- wholesale roaming services,[119]
- satellite services,[120]

Analysis of the S PCS market was thoroughly undertaken by the Commission in 1997 in «the Iridium case»: Iridium had applied for an operational license in Europe and as a consequence the Commission analysed the market as well as the issue of whether there could be policies that would distort competition within the EU and infringe Art 81 (1) of the Treaty.

The decision described the S PCS systems as well as Iridium's strategic investors, which included telecommunications operators, service providers and equipment manufacturers. It analysed further the Iridium network from a technological point of view (segments, operational mode, service distribution). Moreover, it sketched out the relevant product and geographical market, as well as future competitors (Inmarsat, Globalstar). Finally, it assessed legally the possible implications for the competition of the Iridium agreement, concluding that due to the nature of the system, Iridium's partners were not to be considered actual or potential competitors in the S PCS market. Moreover, it concluded that

---

[112] For a thorough presentation, see Commission guidelines on market analysis and the assessment of significant market power under the Community regulatory framework for electronic communications networks and services (2002/C 165/03)

[113] Case IV/M. 856 – British Telecommunications/MCI (II), OJ L 336, 8.12.1997

[114] Case IV/35.337, Atlas, EE L 239 of 19.9.1996, Case IV/35.617, Phoenix/Global/One, OJ L 239 of 19.9.1996.

[115] Case IV/M.975, Albacom/BT/ENI.

[116] Case IV/M.1439, Telia/Telenor, Case COMP/M.1838, BT/Esat.

[117] Case COMP/M.1741, - MCI WorldCom/Sprint

[118] Case COMP/M.1975, Vodafone Airtouch/Mannesmann

[119] Case COMP/M.1863, Vodafone/Airtel

[120] Case IV/350518, Iridium, EE L 16 of 18.1.1997

Iridium would bring competition into the new field of telecommunications mobile market and as a consequence there was no infringement of Art. 81 (1) of the Treaty.

In the decision, S PCS were defined as "systems that provide global, digital, wireless communication services, using constellations of LEO satellites. Services will include voice telephony, paging and basic data services and be provided via portable hand held telephones, vehicle mountain telephones and other subscriber equipment".[121]

On the basis of the decision, S PCS systems were differentiated from any other form of digital, wireless communication means in terms of both infrastructure and services.[122] S PCS systems were specified with regard to performance, technology, service and terminal equipment. As a consequence, they have been described as systems of

- global, digital wireless communications: The function of satellite personal communications systems would be to provide universal or regional communication services, a function which includes terminal mobility provided by wireless access,[123] personal mobility based on personal numbers,[124] and service portability through management of user service profiles;[125]
- composed of LEO constellations;
- to provide voice telephony, paging, basic data services;
- through portable handheld telephones, vehicle mountain telephones and other subscriber equipment;

[121] See Commission Decision of 18 December 1996 relating to a proceeding under Article 85 of the EC Treaty and Article 53 of the EEA Agreement (Case IV/35.518 - Iridium ), OJ L 016, 18/01/1997, p. 0087 ff.

[122] See, for instance, Cases no IV/JV.15 – BT/AT&T, no COMP/M. 1741 – MCI WorldCom/ Sprint on Global telecommunication services, cases no COMP/M.9175 – Vodafone Airtouch/ Mannesmann, no COMP/M. 2016 France Telecom/Orange on seamless pan-European mobile telecommunication services to internationally mobile customers, case COMP/M 1863 – Voda fone/Airtel on wholesale roaming services.

[123] This function refers to the ability of a mobile terminal to access telecom services from any location and to be in motion. The network is able to locate and identify the mobile terminal on the move.

[124] Each user has a unique personal identity (a personal number), and on the basis of this number the network can identify the user as he moves. In this way the users may receive and make calls and access subscriber telecommunications services on any terminal in any location, see John R. Wickens, Who will be the winners? The making of a top player, Pan Asian PCA 96, January 1996.

[125] This capacity refers to the network ability to provide subscribed services. See Raj Pandya, Emerging Mobile and Personal Communication Systems, IEEE Communications Magazine, June 1995.

**A. Relevant Product Market**

a. Technological Differentiation

The significant technological differences between the LEO and MEO as opposed to the GEO constellations leave very little space for substitutability among them. As already pointed out, while presenting the technological differences among them:

- due to system structure, GEO systems require very high ground infrastructure costs;
- it is difficult, if not in a clear line of sight with the satellite, to have a full coverage by the satellite, related to rotation. Geostationary satellites always appear to be at the same point on the horizon, whereas in other orbits, because of the satellites' high speed, earth stations with traceable antennas are needed to track satellites across the sky;
- distance from the earth surface cause significant delays through latency and echo in space. Non geostationary systems, because of their low orbit and the speed of their rotation were seen as the best way to avoid latency and echoes, natural phenomena that annoy phone calls and cause delays of up to half a second in telephone transmissions (as the signal transmitted from a geostationary satellite travels 22. 320 miles to reach the satellite and then to go back to earth);[126]

b. The services provided

The most important services provided by the S PCS have been the following:

- Telephony services, where voice transmission would be the main service. In the telephony services what is very important is the capacity of processing and delivering a voice call within the coverage area of the satellite and with a certain service quality. S PCS voice telephony services aim to be equivalent or better than those of public telephone services;
- Real time data services that is data transmissions that resemble voice communications, the difference lying in their higher integrity;
- Delay insensitive services: to include text messages, electronic mail messages, voice mail, images monitoring data for data collection of any kind of storable data. Examples of the above mentioned services are videotext services and facsimile;
- Paging services, that is short message services;
- Radio determination satellite services. These services would permit the subscriber unit to identify its position to a known accuracy;[127]

---

[126] In particular, as far as delays in round-trip communications are concerned, an echo has a minimum of a 500 m/sec delay, whereas LEOs have less than 30 msec in round-trip communications. See Abbas Jamalipour, as above, p.12.

[127] See very analytical, KPMG, Satellite Communications and their consequences for European telecommunications, Trade and industry, Main Report, vol.I, 1994. See also Emilio

The objective of the operators in the S PCS market would be to provide services in a personal way of communication and to offer a multi-sensorial information interchange. These two elements would mean that these types of services would provide communications in any place, at any time, without depending on a specific operator, seamless and unrestricted to real-time voice telephony.[128]These capabilities would presuppose a series of services as well as technical and operational requirements for the provision of the services. These would include service ubiquity, a unique personal identification number (upin), personal and terminal mobility, universal service provision (service requirements), full networks and services interconnection, inter-operation, management of users' location and mobility and behaviour (technical and operational requirements).[129]

c. Mobile terrestrial digital complementarity

The S PCS have been designed as a user oriented communications system, which would provide ubiquitous services to users who did not have a fixed geographical location. Moreover, they aimed to provide mobile services outside the coverage of mobile terrestrial systems in order to meet the needs of customers in both rural and densely populated areas. As a consequence, S PCS systems would complement the terrestrial ones (networks that had a 100% coverage in the urban areas, but very low in rural parts, desserts, mountains and areas of high latitudes), with the aim of being more attractive in terms of cost, quality, signal strength, appearance and attractiveness of the handsets.[130]

Various scenarios of market evolution were drawn, from slow S PCS market development scenarios to fast development ones. However, even the slowest scenarios could not predict the very low percentage of penetration of the systems. Slow development scenarios foresaw a market of 1.3m subscribers, whereas high development ones moved up to an estimation of 14m subscribers. The most positive scenarios foresaw a market of even 30m users by the end of 2005. It was unlikely that the S PCS systems would penetrate Europe to the same extent as the mobile terrestrial, which had reached such a significant number of subsc-

---

Lera, Towards multiservices personal communications: perspectives for a sector structure evolution, in Telecommunications policy, vol. 20, No. 7, 1996, pp. 481-496.

[128] Ibid., p. 482.

[129] Service ubiquity means the ability to obtain the same service features in any ambit, be that home, office, district, town, or rural area. UPIN is the user identity recognized and authenticated by any services provision network and publicly communicated according to users' instructions. Interoperation between terminals and networks allows the connection between all types of terminals and all networks. Location means the point where the user has for the last time identified him/herself and behavior is related to the behavior of the user towards the concrete terminal and means that the services offered and the connectivity should not be affected if the user changes terminal. For more details see Emilio Lera, as supra.

[130] These are the GSM, DCS 1800, DECT, PSTN.

ribers, despite the enthusiasm and hope of the S PCS designers and investors. Many users from outside Europe, with access to different cellular standards, were expected to use the satellite function of dual mode phones[131] in Europe. In particular, the demand for satellite PCS in Western Europe was expected to reach less than 200,000 subscribers by the year 2010. In Eastern Europe, due to the absence of terrestrial alternatives, the demand was expected to reach approximately one million users by the year 2007. However, on a global scale, demand has been disappointing for all systems that became operational. Globalstar, to give but one example, ended up having 83,000 customers, after the first year of operation, instead of the millions they had initially predicted.

d. User groups

The identified critical market areas, at which the systems have been aiming, have been various. First, global roamers both private and public sector, which demanded the availability of global communications. Second, users in territories that could not justify terrestrial network expansion, due to local subscriber density or to the expense of additional infrastructure. Third, users without a fixed service territory. Fourth, rural and/or developed areas in which a few priority circuits could vastly improve telecomms and general economic welfare.[132]

**B. Relevant Geographical Market**

S PCS systems have emerged either as global or as regional. Iridium, for instance, provided global services. The Commission has left the definition of the geographical market open, explaining that its exact scope is difficult to ascertain.

*IV. Economic Actors*

**1. S PCS consortia and the formation of Global Economic Networks (GEN)**

The main feature of the S PCS economic landscape has been the formation of strategic alliances, mainly in the form of consortia,[133] which has constituted part of the phenomenon of the emergence of GEN during the last decade.[134]

---

[131] Dual mode phones would function as a terrestrial mobile phone, like GSM, or DECT as well as a satellite phone, operated by Iridium, Globalstar, Inmarsat or other S PCS system).

[132] Mobile via satellite: bent pipe dreams? PNE, December 1997, p.45.

[133] The term will be further in the chapter chapter explained.

[134] In the past two decades important events have been taking place on the global economic scene. A very fast internationalisation process, an increased multipolarity and multidimensionality in economic structure in the international economic relations, have made finance markets, trade, investment, product markets and national transactions acquire a global dimension. See Dalibor, Eva, Handelpolitische Mutlilateralismus, p.10 ff. The author also deals with concepts such as "integration", "convergence", "worldwide economy" and "internationalisation", as levels and parts of globalization.

**A. Global Economic Networks (GENs)**

GENs have been a recent phenomenon of globalisation. They have caused global economic relations to be characterised by the worldwide expansion of industries and the emergence of regional trading blocks.[135]

By definition a network denotes a system or group whose members are connected in some way, vertically or horizontally.[136]The global economic networks have been broadly defined:

*"cross-national production networks involve the organisation across national borders or research and development activities, procurement, distribution, product definition and design, manufacturing and support service, their basic features including fragmentation of the production, the dispersion of production facilities, world-wide sourcing and intrafirm trade"*[137]

Global economic networks reveal chains of economic actors that are networking in terms of production, manufacturing, designing or providing a service throughout the globe. These actors have often managed, through concentrated economic power, to bypass national governments and to influence their policies.[138]This strategy of networking exercised by the global economic networks has made it possible for the actors to integrate a variety of competencies among industries which have been affected by the merging of electronics, informatics, telecommunications services and equipment technologies.[139] GENs emerge from a certain strategic behaviour, having particular locus of power and a specific hierarchy.[140]However, they presuppose a vertically and horizontally integrated industry, with many stages of supply and production and many different actors participating. The telecommunications market has been a representative example of GEN formation because it has been a market where the increasing importance

---

[135] Marc Humbert (ed.), The impact of globalisation on Europe's firms and industries, 1993, p.115 ff. For a general approach on the impact of globalisation of economy, see Horváth\ Merzrath, Globalisierung, Springer, 1997, p. 67 ff.

[136] See Mansell Robin, Information, Organisation and Competitivenesss, Networking strategies in the 1990s, in Antonelli C., (ed.)The economics of information networks, Elsevier Publishers, 1992.

[137] See Francis Snyder, Globalisation and Europeanisation as friends and rivals: European Union Law in global economic networks, EUI, working paper, no 8, 1999. p. 9. Snyder concentrates on the analysis of international production networks , see. p. 9 ff.

[138] See also Michael Hodges, Multinational corporations and national governments, 1974, p.4.

[139] See Antonelli, C., Technological change and multinational growth in international telecommunications services, in Noam/Wolfson, (ed.), Globalism and Localism in Telecommunications, Amsterdam, Elsevier, 1997, p. 283.

[140] See Francis Snyder, Governing Economic Globalisation: Global legal Pluralism and European Law, p. 340.

of transnational interaction has revealed the power of the economic agents of interaction, which is the industry.

A good example of this transformation would be the telecommunications market: from the stable and predictable structure of the telecommunications industry in the seventies, as directed by international cartels of national governments, the industrialists have taken the lead forming global networks of supply and production. In this way, the common consensus that the global economy has become not only more complex, but also more tightly interconnected, as more and more industries are oriented towards a global context, instead of local, regional or national, has been reaffirmed.[141]In this context, the state has become dependent on other non-state actors for issues related to the installation and operation of networks or dealing with their externalities, such as processing, data flow and use.[142]Moreover, two decades ago, companies would strike deals with host governments to bring in resources and skills in exchange for privileged access to local markets. Now motives have been shaped differently. Motives are shaped not by reasons related to privileges granted by the state, but by the desire and need to identify, penetrate and serve market segments and track useful resources. Government-funded subsidies are becoming irrelevant as a criterion for the decision of a firm to act in a local market.[143]

**B. Strategic Alliances and S PCS**

Within the general tendency for the formation of GENs, the S PCS market has attracted the interest of many industrialists with a dynamic presence in the telecommunications sector, not only sectoral and regional, but also global power. As a consequence, various industrial players have been involved. In particular, at an investors' level the strategic alliances were formed between the main industrial players. Enterprises with deep pockets and power teamed up forming coalitions in form of consortia.

---

[141] See Dicken Peter, Global shift: industrial change in a turbulent world, 1988, Paul Chapman Publishing Ltd, p.3, Pekka Tarjane, The limits of national sovereignty: issues for the governance of international telecom in William H. Melody (ed.), Telecom reform, principles, policies nad regulatory practices, 1997.

[142] To mention but a few examples, Ericsson has been responsible for the installment of the GSM network in China, Nokia has put the first WAP (Wireless Application Protocol) on the market with videoconferencing possibilities for GSM networks in markets such as the German and the Finish one. Moreover, Ericsson developed WAP in Malaysia and Morocco.

[143] See Kenichi Ohmae, The end of the national state, The rise of regional economies, HarperCollins Publishers, 1995, p.3.

a. Strategic alliances: The concept

Technological progress, the globalisation of the telecommunications markets together with the presence of extremely large economies have made national markets too small to support the level of R&D required to develop new technical equipment and to provide new advanced services.[144] Furthermore, on the demand side, big multinational customers have new needs, requiring global and seamless services, with a consistent service level 24 hours a day, a level which should not be affected by geographic location, national borders, languages, currencies, time zones, or the inadequacies of local infrastructures.[145]

Strategic alliances started spreading especially in the 1990s, at a very rapid pace.[146] They have been transnational partnerships that build strong economic networks, and that have as their main feature mobility and dynamism.[147] These partnerships have been formed either between firms, or between firms and national telecommunications organisations -which after having lost privileges[148] started looking for new markets, restructuring and modernisation-or between national telecommunications organisations and other industries, seeking profit.[149]

---

[144] Jerry A. Hausmann, Joint Ventures, Strategic Alliances and Collaboration in Telecommunications, Regulation, v.14, 1, Winter 1991, p.1, Weidenbaum Murray L., When Business cross international borders, Transnational Lawyer, vol.7, 1994, p.129-132.

[145] Miguel Angel Pena Costelot, The application of competition rules in the telecommunications sector: Strategic Alliances, Competition Policy Newsletter, number 4, v.1, Spring 1995. Also, for the reasons of the appearance of strategic alliances and multinationals in general, see Horváth/Merzrath, Globalisierung, 1997 and Luccatelli, The making of transnational regimes, Zürich Universitaet, Diss., 1996, p. 17.

[146] There is also the very critical opinion on the strategic alliances, which argues that the consumers are tired of hearing about so much about big industry names and of not seeing satisfactory results. This means that they want to see their capacities and how much they can offer to world communications. If they do not do it, the consumers will loose their faith, and their future will not be that bright. See Public Network Europe, June 1999.

[147] See Peter Lorance and Johan Roos, Strategic Alliances, formation, implementation and evolution, 1992.

[148] In the framework of liberalisation and deregulation process, the NTOs had their functional and operational privileges separated and remained units more commercially oriented. This separation took place with the Services Directive.

[149] Meni Stuliadou, The Application of Competition Rules in the Telecommunications Sector, Telecommunications Policy, 21(1), 1997. Grewlich categorizes two types of strategic alliances. The first category comprises of alliances with partners that are new entrants in the telecommunications market and that want to achieve favorable synergy-costs effects at the time of market entry. The second type comprises of alliances between already established telecommunications organisations, which have either a global or a regional expansion. In the first case, there are alliances like Cable & Wireless-VEBA, International Satellite Partners, whereas in the second BT-MCI, ATLAS, Global One. Grewlich, Konflikt und Ordnung in der globalen Kommunikation, 1997, p. 108 ff.

A strategic alliance has been defined as:[150]
«a bilateral or multilateral relationship characterised by the commitment of two or more partner firms to a common goal. A strategic alliance typically includes a constellation of arrangements involving a) technology swaps b) joint research and development or co-development, and / or c) the sharing of complementary assets, such as where one party does manufacturing and the other distribution for a co-developed product.[151]»

Strategic alliances have various objectives, such as penetrating quickly a certain market, or developing long term strategies aiming at world leadership.[152]This interaction has been adapted to the model of a market-based economy, and determining new models of control and power, defined mainly by the capacity to exploit advanced technology.

The specificity of strategic alliances is that they constitute a transnational phenomenon, which involves multi-dimensional and multisectoral patterns of interaction as well as mutual sensitivity.[153]The reasons for entering an alliance are

---

[150] The definition has been elaborated by Thomas M.Jorde and David J.Teece, Innovation, Cooperation and Antitrust, in Antitrust, Innovation and Competitiveness 55, 1992.

[151] The authors, supra, explain furthermore that, "a strategic alliance can never have one side receiving alone, as it is not a unilateral exchange transaction. Strategic alliances do not include mergers because alliances by definition cannot involve acquisition of another firm assets or controlling of another firms stock. Alliances do not need to involve equity swaps or equity investments, although they often do so. Strategic alliances without equity typically consists of contracts between or among partner firms that are non affiliated."

[152] Also Karl-Heinz Neumann, Interconnection Agreements in Telecommunications, Wissenschaftliches Institute, Bad Honnef, Working Paper, 72, 1991. The objectives of partering in horizontal and vertical alliances vary. As regards the horizontal alliances, they try though the alliance to survive in the global market by gaining the maximum efficient scale, putting up with the high costs of the new technologies, facing the uncertain demand of the new services and the uncertainty of evaluation of resource protection, and by establishing international networks. On the other hand, vertical alliances also try to put up with the uncertainty of the new market, to achieve a better market positioning, to obtain new skills, to control the channels open to the customer, to move into higher value areas, to overcome market barriers, to possess technology, and to protect of resources.In many cases companies have tried through alliances to increase their profitability and competitively with the leading market players, like in the case of Cable and Wireless Communications. See Jong-Geun Oh, Global Strategic Alliances in the telecommunications industry, Telecommunications Policy, vol. 20, 9, 1996, p.713 ff. Wrona, Globalisierung und Strategien der vertikalen Integration, Wiesbaden, Gabler, 1999, p. 9-29.

[153] The phenomenon of strategic alliances is considered to be one of the most important phenomena challenging approaches in strategic management since it breaks the fragile limits between competition and co-operation and it requires new analytical tools to understand the nature of these agreements. See Bernard Garrette, Bertrand Quelin, An empirical study of hybrid forms of governance structure: the case of the telecommunication equipment industry, Re-

various: to find embryonic technology, strategic market segments, or new business opportunities. Moreover, globalisation has allowed strategic alliances to exploit three important advantages. First, to identify and penetrate new markets, seek new partners, exploit their experience and achieve substantial savings by using the partners' distribution networks. Second, to gain more profit, as globalisation allows production in low cost countries. Third, to gain economies of scope, as broad-based companies may gain resources over local niche players and, through lower production costs, gain competitive advantage in the global market.[154]

b. The appearance of the alliances

The partnership of firms and other entities, is a relatively new phenomenon, which started in the eighties when companies realised that it was impossible for a single company to possess all the resources needed to deal with the complexity and the expenses of modern technological development.[155]After the Second World War, and until the 1970s, it was the US that dominated the trade of technology equipment and of services. The main features of the American firms were autonomy and absence in co-operation between them. This was because each one of them developed individual strategies for approaching European or Japanese firms. Since the late 1970s there have been certain patterns of interaction related to the technological competitive advantage of the three super powers, the EU, Japan and the US, as each of these powers presented a different competitive advantage, which the others did not have. Japan has been very strong in microelectronics and robotics, the US in space and satellites and the European Union in mobile communications. As a consequence, the patterns of interaction have been taking place in the so-called triads: US-EC, EC-Japan and US-Japan.[156]These patterns of interaction have, in the last years, been multiplied through the increase of foreign direct investment (FDI). This has especially been the case in Europe. The FDI provided the means to penetrate markets that had traditionally been protected, and which were to be liberalised.[157]Penetration of

search Policy 23, 1994, 398. See Harvard Business Review(ed.), Global Strategies, Insights from the world's leading thinkers, 1994, 7th Ed., Harvard Business School Publishing, p.110. However, strategic alliances comprise of a rather ambiguous factor, which is the different integrated elements and their location in various cultural, economic, political and social environments. See Harvard Business Review, as supra, fn. 398, Colin Long, Stephen O. Spinks, Alliances, Mergers and Acquisitions in Telecommunications, International Telecoms Review 1997, to be found in www.coudert.com/teleacq.htm.

[155] See J Monlouis, the future of telecommunications operators alliances, Telecommunications policy, vol. 22, 1998, v.8, pp. 635-641.

[156] The number of partnerships among American and Japanese firms was the smallest. Both US companies and Japanese companies preferred as partners companies for Europe.

[157] See Nunnekamp/Gundlach/Agarwal, (Ed.), Globalisation of Production and Markets, Tübingen, Mohr, 1994; Ostry, Beyond the border: the new international policy arena, in

the European markets was easy for the Japanese manufacturing firms (NEC, Toshiba, Panasonic, Mitsubishi), which preferred to penetrate the European market than the US market. As a consequence, many scholars talk about triadisation instead of globalisation.[158]

In the early 1980s the world was dominated by various actors from the triads in specific sectors.[159]In electronic equipment: AT&T, GTE Harris, IBM, ITT Mitel, Northern Telecom (US), Hitachi, Fujitsu, NEC (Japan), Alcatel, Ericsson, GEC, Phillips, Siemens, Thomson (Europe), in telephone equipment: AT&T-Phillips, GTE-Italtel, ITT (US), Fujistu, NEC (Japan), CIT-Alcatel, Ericsson, Siemens, Thomson, Plessey (Europe), in overseas transmissions: AT&T, ITT, MCI-Western union, RCS, TRT (US), Cable&Wireless, France Cables&Radio, Itacable (Europe), Satellite carriers Intelsat, Orion.[160]As a consequence, in the first half of the 1980s, the international market was dominated by large national groups. In the US there were AT&T, IBM, Motorola; in Japan NTT and the supplier firms, OKI, Hitachi, Futjitsu, NEC; in Europe German and Scandinavian firms: MBB, AEG, Siemens, Bosch, Nokia, Ericsson. It was American firms that first started developing international strategies. The first alliance was formed by AT&T with Phillips and Olivetti in the computer market in 1983.[161]

During the second half of the 1980s the scenery started to change. Partnerships started to increase as mobile communications, digital transmissions and telematics started being an item for collaboration. However the structure of networks still remained the same. They were based on national-regional groups, such as,

---

Kantzenbach/Scharrer/Waverman (ed.), Competition Policy in an independent world economy, Baden-Baden: Nomos, 1993 p. 261. In particular Sylvia Ostry categorizes the growth of linkages in international trade in three phases: the first focuses on trade expansion during the 50s-60s, the second distinguishes the structural characteristics of the financial integration of the 70s-80s and the third the phase of globalisation.

[158] See Petrella, Globalization and internationalization, in Boyer/Drache, States against Markets, p.77.

[159] On the expansion of multinationals in the eighties, with particular reference to the US case, see Vernon, Beyond Globalism, Remaking of American Foreign Economic Policy, The free press, NY, 1989, p. 109-112.

[160] For a very good illustration of the situation in the telecommunications sector in the early eighties see International business week, telecommunications, the global battle, special report, October 24,1983. p.62.

[161] In late 1983, AT&T and Olivetti entered an alliance in which AT&T purchased a 25% equity stake in Olivetti for $260 million. The agreement enabled AT&T to benefit from Olivetti's office products in the US. Merrill Brown, AT&T to buy 25% of Olivetti, Wash. Post, Dec. 22, 1983. Additionally, the alliance called for the joint development of products, cross-licensing of selected hardware lines for domestic manufacturing, and access for Olivetti to AT&T's Bell Laboratories' research and development product.

NEC, Fujitsu in Japan, Bosch in Europe and Baby Bells in the US.[162]However as the firms realised the benefits of entering into such partnerships, the number of alliances started to increase. First, as regards R&D, international partnerships could mean, for a single firm the reduction of costs and a reduction in the company's financial risks. Second, the innovation process could be accelerated; and third, as far as issues related to market access were concerned, penetration of foreign markets could be facilitated, and the range of new products promoted could be expanded.

Globalised markets started to require a complex mixture of product, price, promotion and distribution. They required new approaches to ownership overseas and the development of new types of relations.[163]The complexity of the global environment and its uncertainty -caused by the lack of technology and knowhow- has made the number of alliances rise revealing three important dimensions of global economic relations.[164]First, the increased in globalisation itself, as composed of science and technology, markets and economies, strategies and institutional forms.[165]Second, the general socio-political context and in particular the changes in the roles that nations, regions and societal groups play. Third, the uncertainty caused by information technologies and the phenomenon of convergence.

As international economies were fragmented, it was the strategic actors that presented themselves as the primary agents of economic development.[166]This became the major driving force of deregulation. Globalisation made the presence and the dominance of strategic actors necessary and unavoidable whereas state and interstate systems served mainly to ensure that market logic still worked. The use of new resources has redefined the basis on which global economic relations are conceptualised, from international co-operation to interactive relati-

[162] See Karl-Hainz Neumann, Interconnection Agreements in Telecommunications, Bad Honnef Wissenschaftliches Institute für Kommunikation, 1993.

[163] Jong-Geun Oh, Global strategic alliances in the telecommunications industry, in Telecommunications Policy, vol. 20, n. 9 , 1996, pp. 713-720.

[164] See Nanette S. Levinson, Stratégies interorganisationnelles dans les technologies nouvelles: alliances transnationales et mondialisation, in Abdelaziz Mouline(ed.), Les Alliances stratégiques dans les technologies de l'information, 1996, p. 19 ff. Also on the rise of a specific type of alliances, the so-called consortia, see Richard Hawkins, The rise of consortia in the information and communications technology industries: emerging implications for policy, in Telecommunications policy, 23, 1999, pp. 159-173.

[165] Nanette S. Levinson, p. 18.

[166] Robert W. Cox, A perspective on Globalization, pp. 21 ff., in Globalisation critical reflections, edited by James H. Mittelman, Globalisation: Critical Reflections, 1996.

ons based on international networks of communications.[167]Furthermore, the most paradoxical observation has been that the movement towards liberalisation has resulted in a further concentration of industrialists, especially amongst national champions (f.e., GTE, Alcatel, Siemens, AT&T and Phillips, Matra, Ericsson and CGCT, OTE and KPN).The economics of a global strategy of alliances lie in the dynamic of network externalities: the larger the network, the greater the benefit of belonging to it.[168]

c. Strategic alliances in the S PCS market

In the last ten years an important number of strategic alliances have made their appearance. Technological progress in the satellite sector and the increased demand for the provision of huge amounts of data all over the world, combined with the advantages that the new technology could offer,[169]have attracted the interest of industrialists and telecommunications organisations, who have teamed up and presented many global satellite projects.[170]

Many of the projects that appeared have focused on the operation of the new Big and Little LEO systems for the needs of mobile communications (Iridium, Globalstar, ICO, Orbcomm) as well as systems based on broadband satellite constellations for internet-based multimedia applications (Teledesic, Skybridge, Horizons).[171]Most of the projects have been operated by strategic alliances in a form of a consortium.[172]

A consortium has been defined as an informal alliance of firms, organisations, and (sometimes) individuals that is financed by membership fees for the purpose of co-ordinating technological and market development activities. Informality is a key characteristic as most conventional industrial networks operate through formal sub-contracting arrangements or through partnerships and join ventures.

---

[167] Saskia Sassen, the spatial organisation of information industries: implications for the role of the state. in James H. Mittelman, Globalisation Critical Reflections, 1996, OECD, Towards a global information society, 1997, p. 7.

[168] Antonelli in Noam, Globalism and Localism in Telecommunications, p. 283.

[169] Reduced costs of ground infrastructure, compared to the costs of geostationary satellites, ability to offer services of voice telephony and in the future multimedia applications, global coverage, allowing the systems to provide services comparable to those of landmobile terrestrial services.

[170] In particular, in 1994-95 there were more than 55 filings by the ITU for S PCS systems.

[171] In official documents they are considered in broad terms to be S PCS, however their legal status is not clear. See below, chapter IV, in the S PCS decision.

[172]This is a definition on consortium given by R. Hawkins. For more information on consortia (advantages, disadvantages, origins), see R. Hawkins, The rise of consortia in the information and communications technology industries: emerging implications for policy, Telecommunications Policy, 23 (1999) 159-173.

Typically, consortia set out very explicit objectives and agendas, but pursue them through very "informal" working procedures.

Satellite manufacturers, like for instance Hughes or Loral, were not the only ones interested in the S PCS. There has been an increased interest of economic players, including: mobile providers (STET Italy, Vodafone), aerospace industry (Deutsche Aerospace, Aerospatial)), telephone providers (Bell Canada), NTOs (France Telecom), space industries (Orbital Sciences), mobile and microchip manufactures (Motorola, Qualcomm, Alcatel), industries active in the telecomm sectors in general( NEC, Mitsubishi, Hyundai). All these interested players have started teaming up in all possible combinations: satellite manufacturers with mobile manufacturers and aerospace industries; space industries with telecommunications organisations; mobile providers with mobile manufacturers and telephone providers.

Strategic alliances in the satellite sector had to face the following particular specificities.

- The S PCs network technology was untested. The network had not been previously operated, in particular for LEO constellations with intersatellite links. As a consequence, the risks were big, as many technical difficulties, especially related to space segment and the orbital positions, had to be taken under consideration. Its success would only be proven when deployed fully in operational configuration and through handling large volumes of traffic, something which would only happen in time.[173]
- Strategies had to be developed in order to attract investors and services providers who would distribute and market their systems. The nature of the services offered would provide extended regional or global coverage. As a consequence, big chains of vertical and horizontal integration had been developed, and the investment needed reached far beyond initial expectations.

### *V. S PCS alliances*

### 1. Global players - Big Leos

#### A. Iridium

The Iridium idea, a constellation of small satellites that would allow calls anywhere on the globe, was established in 1987.[174]In 1990 the system, mainly backed by Motorola, filed for license by FCC, and in 1992 it was granted an experimental license to construct and launch satellites. The company tried to find investors and in 1993 it had already raised as first capital $800 million, of the $3.5 (billion?) which had been the initial estimation of the cost of the systems.

---

[173] Case IV/350518, Iridium, EE L 16 of 18.1.1997, p.7

[174] Motorola Satellite Communications is in charge of spacecraft construction.

Iridium was backed by 19 powerful investors, including mobile manufacturers, satellite constructors, mobile service providers, and various industries involved in the telecommunications sector.[175]

Iridium started as the first global satellite system on the market. It begun its operation on the 1rst November 1998, after a two month delay due to technical difficulties and delays in the delivery of the equipment. It aimed to attract millions of customers, however it ended up with a few thousand. As a consequence, given the high operational costs, the need for constant investment and the lack of revenues, Iridium was forced on August 1999 to file for Chapter 11, against bankruptcy protection. Since then, Iridium has tried to raise fund that would permit its further operation. In the year 2000, as no buyer was found, the system stopped providing services and decided to start pulling satellites back to earth, at that time there were only 55.000 customers (500.000 customers were already expected in 1999). However, Iridium's assets were bought at an auction for $25 million in December 2000 by Iridium Satellite LLC, a privately held corporation. Iridium has since then been providing services to the US government.

| | |
|---|---|
| 1987 | « Birth » of Iridium |
| 1990 | Filing for license at FCC |
| 1992 | Granting of experimental for satellite construction |
| 1993 | 1rst Capital raising |
| 1998 | Beginning of operation |
| 1999 | Filing for chapter 11 |
| 2000 | Purchase of Assets by Iridium Satellite LLC |

**B. Globalstar**

Globalstar, the second Big LEO constellation project, was designed in the late eighties. In 1986 Qualcomm began to work on a world-wide satellite-based telecommunications service, in order to offer wireless telephone and other telecommunications services world-wide. In 1991 Loral Qualcomm Satellite Services was formed for the purpose of filing an application for a construction permit for

[175] Motorola, Iridium Africa Corporation, Iridium Andes Caribe, Iridium Brasil, Iridium Canada, Inc., Iridium China LtD., Iridium India Telecom Limited, Iridium Italy, S.P.A., Iridium Middle East Corporation, Khrunichev Enterprise (launchers), Lockheed Martin Corporation, Nippon Iridium, Kyocera Corp., O.TEL.O, Pacific Iridium Telecommunications Corporation, Raytheon Company, SK Telecom, South Pacific Iridium Holding, Sprint Iridium, Thai Satellite Telecommunications Co.

a Globalstar type system to which the construction, launching and operating license was granted in 1995.

Globalstar, besides Loral and Qualcomm, has attracted the interest of various strategic partners, which joined the project in 1994.[176] Among them, there were Alcatel, a very big French aerospace and telecommunications equipment manufacturer, France Telecom, Alenia, an Italian aerospace manufacturer, Qualcomm Aerospace, Hyundai, Dacom (Korea's second telecom carrier), Air Touch, and Vodaphone, a British cellular radio operator. The system became the second one to provide, since January 9th, 2000 S PCS services, initially on a limited basis (US zone), and later through Europe.[177] However, in early 2001 it ended up in debt and in February 2002 it filed for bankruptcy protection for Chapter 11. Early 2003 an investment group called new Valley Corp promised to invest $55 million in order to help the company complete the Chapter 11 restructuring process.

**Globalstar**

| | |
|---|---|
| 1991 | Formation of the consortium for the construction of the SPCS system |
| 1995 | Filing for application at FCC |
| 2000 | Beginning of service provision |
| 2002 | Filing for Chapter 11 bankruptcy protection |
| 2003 | New Investor found-Restructure |

## C. ICO

ICO was founded in January 1995 as a MEO system, which intended to start providing satellite personal communications in early 2000. It raised the initial capital of $1,4 billion, -of the $4bn of total cost very quickly and attracted important investors from Europe, US and Japan, like Alcatel, British Telecom, Inmarsat, COMSAT, TRW, NEC.[178]

---

[176] Loral Corporation is a leading defense electronics company and has entered the satellite industry through its acquisition of Ford Aerospace in 1990. In 1991 Loral sold 49% of Loral/Space systems to Aerospaciale, Alcatel Espace and Alenia. The combined resources of this group in R&D, marketing and production. make Loral/Space Systems a significant player of the satellite market. Qualcomm was established in 1985 and begun commercial activities in 1989 by using leased capacity on Ku-Band satellites.

[177] On February 2nd, 2000 Globalstar started providing services in nine European countries and plans to launch in other European and non-European as well. See Vanessa Clark, Globalstar Launches European Service, www.totaltelel.com, 2 February 2000.

[178] In particular ICO's investors were, among others, the following: Arabsat, Bahrain Telecommunications Company, Beijing Marine Communication and Navigation Company, BT,

ICO started as the S PCS Inmarsat's plan, designed to provide voice, data, fax, paging, messaging and position services. For the supply of its products to maritime and aeronautical customers, ICO counted on its former parent INMARSAT. It had roaming agreements with cellular operators in Europe and in Asia. However, due to fund raising difficulties ICO filed for Chapter 11 (Friday 27 August 1999) before even becoming operational. However, capital has been found. McCaw of Teledesic announced his intention to help the system overcome the crisis and through funding the new ICO has emerged after the acquisition in May 2000 of ICO by Teledesic's investor.[179]ICO has been in the process of restructuring, having been oriented towards internet-based applications.

| | |
|---|---|
| 1995 | Foundation of ICO |
| 1999 | Filing for chapter 11 |
| 2000 | Acquisition of ICO by Teledesic |

**2. Regional S PCS**

Besides the global Big Leo systems there have also been regional S PC projects that focused on certain regions. Many made their appearance, however most of them remained stagnant mainly due to financial obstacles and were either withdrawn, stopped or integrated into other projects. The entrants, which were most likely to succeed were ECCO and Ellipso. ECCO has been formed in 1991 in order to provide global MSS services and emerged after Constellation Communications International, owned by Orbital Sciences Corporation, Bell Atlantic Global Wireless, Raytheon E-Systems and SpaceVest, were joined by the Brasilian PTT (Telebras) and the Brazilian Space Research Organisation (INPE). It has been a 12 LEO equatorial constellation with regional coverage with plans to be operational in 2002.

Ellipso has been another regional project with investors Westinghouse, Harris Corporations and Israeli Aircraft (Former Mobile Communications Holding INC) to which the launching and operation's license was granted in July 1997.

Bureau of Maritime Affairs (Liberia), Compania Portuguesa Radio Marconi, Comsat Argentina, Comsat (US), CS Limited (Thailand), Cyprus Telecommunications Authority, Empresa National de Telecommunicaciones (Columbia), Empresa National de Telecomunicaciones de Cuba, Etisalat (UAE), General Post and Telecom Company of Libya, General Telecommunications Organisation (Oman), Hellenic Telecommunications Organisations, Hughes Telecommunications and Space Company, ICO Korea, INMARSAT, International Communications Organisation Cameroon, Lebanese Ministry of Posts and Telecommunications, Kuwait Investment Authority, Mauritius Telecom LTD, Israel Ministry of Communications, Ministry of Posts, Telegraphs and Telephones and Saudi Public Investment Bulgaria NEC Corporation (Japan), Nigerian Telecommunications Limited, Pakistan Telecommunications Corporation, Office des Postes et Telecommunications (Benin).

[179] See http://www.ee.surrey.ac.uk/L.Wood/constellations/ICO.html.

Ellipso was expected to be, at first, regional, covering the equatorial countries (Asia, Africa, Latin America) and later hoped to offer followed by global services. However, due to financial drawbacks in March 2001 Ellipso reached an agreement with ICO-Teledesic Global Limited.

Besides these projects there have been many others that had announced their operation at a regional level. Regional entrants included not only LEO and MEO systems, but also many GEO systems, like Thuraya. Aces (Asia Cellular Satellite), which was considered to be the most advanced regional GEO player, has been a $900 million, project of two GEOs(one operational and one back up), aiming to provide coverage for the Asia Pacific Region.[180] Africom has been a $650 mn. project by Africom and Lockheed Martin of one GEO, which aimed at covering Africa. APMT is a another regional project of $900 ml, initiated by China Satellite Launch&Tracking Control general, China United Telecommunications Satellite CO, China Telecommunications Broadcast Satellite, China Overseas Space Development&Investment Co. and Singapore Technologies. This project targeted China, Philippines, Vietnam and South East Asia. From these projects Thuraya has been providing services, mainly most of the other systems remained stagnant due to financial difficulties.

**3. Little LEOs: Orbcomm**

The main active player in personal communications messaging, monitoring and tracking services has been Orbcomm, a little LEO satellite constellation of $500 million.[181] Orbcomm was the first Little Leo system, consisting of 30 satellites, which entered the market almost at the same time as Iridium. It was the joint venture of Orbital Sciences Corporation (space technology company), Teleglobal of Canada (intercontinental telecoms industry), and Technology Resources Industries Bad. of Malaysia (cellular operator). The system has been similar to a two-way paging or e-mail. It had 16 regional distribution partners

---

[180] Partners in the project have been Pacific Satellite Nasantara of Indonesia, Philippine Long Distance Company and jasmine international PLC of Thailand. See Ovum, LEOs, MEOs and GEOs: the Market Opportunity for Mobile Satellite Services, 1998, p. 225.

[181] Orbcomm has been a joint venture of Orbital Sciences Corporation, Teleglobe Inc. of Canada and Technology resources Industries BhD. Of Malaysia. Orbcomm used frequencies of 137-138 MHz and 400 MHz for the downlink as well s 148-150 MHz for uplink. The constellation begun with 28 satellites: then with additional permission it grew to 36 and on April, OCC (Orbital Communications Corporation), an affiliate of Orbcomm, bas been authorised by the FCC to launch and operate 12 additional satellites. Orbcomm has been granted an operations license in Europe and has teamed with Orbcomm Europe, a consortium led by Telespazio, Stet, OHB of Germany, Satcom of UK and the Swedish space corporation of Sweden. For more see www.orbcomm.com.

and provided services in 105 countries.[182] Orbcomm provided its services in Europe through Orbcomm Europe, a consortium of European telecommunication companies, led by Telespazio, OHB of Germany, SATCOM of UK and Swedish Space Corporation. However, Orbcomm faced severe financial difficulties and on 15 September 2000 filed for Chapter 11 against bankruptcy protection. The $800 million Little Leo was sold through an auction for $16 million. Moreover, there were other Little Leo Systems, besides Orbcomm, planning to enter the market,[183] like Safir, a German project targeting the European market, Iris, a Belgian project, aiming to provide services for European transport companies, and E-Sat, a US led project for oil and gas companies, which all stopped due to financial constraints.[184]

### 4. Giga LEOs

Besides the Big and Little LEOs, industrialists have also invested heavily various Giga LEO projects. The specificity of these projects would be their increased capacity that would allow them to transmit a larger amount of data in a shorter time, and to be adequate for multimedia applications. Industrialists involved have been very ambitious and expected a lot of profit, however there have been many sceptical analysts, who saw the advanced technology as too risky. The main actors were expected to be the following systems:

#### A. Teledesic

The company was founded in 1990 and in 1994 the initial system design was completed. In 1997 the FCC license was granted and frequencies were, during the WARC 97, allocated for individual use (designation of 500 MHz). Teledesic

---

182 Among its partners: ALP Technologies, American Millennium Corporation, ARINC, Axiom Engineering Limited, BANLEN, Beta Corporation, Blocke Communications, Caribbean Technologies, Intersat, Infosat Telecommunications e.t.c. Each of these partners had a certain exclusive area.

183 See the two ERC decisions, ERC/DEC/(99)05 and ERC/DEC/(99)06 on S PCS <1 Ghz., where these systems are mentioned as strongest candidates.

184 Iris has been Belgian aproject (involving Sait Belgium, ESA, OHB), which targeted low cost store and forward services for European transportation companies, having a large number of vehicles spread over isolated areas in Europe and especially Africa. The project aimed to provide trucking, shipping, services to professional teams in remote areas, (road or damage building, railway maintenance, pollution control missions), international travellers, e-mail. Terminals in many cases would be a kind of satellite modem, linked to the users' personal computer or to an automatic data collection device. E-Sat (ex Starsys) has been a US project formed by Echostar Communications Corp.(80%) and DBS industries Corp. (20%) It has been a six satellite constellation of $50 million, which would target primarily the oil and gas markets for store and forward services and it would also support services related to direct-to-home broadcasting.

has been the early leader in identifying the internet market for NGSO and it was one of the first systems that filed for the FCC Ka-Band license.[185]

The project has been estimated at a cost of $9 billion. Among its investors, Motorola, the Boeing Company and Matra Marconi joined efforts.[186]Motorola led the industrial design of the system and has contributed $750 million in cash and development assets, taking 26% of the stake.[187]The system was scheduled to operate in the year 2003. Teledesic aimed to reach 20 million subscribers and $6,5 bn revenues in the year 2006. The concept was to serve remote areas, bypassing the information society and being an "internet in the sky". Teledesic has been seeking alliances with service providers in countries all over the world, the starting year of operation has been postponed until 2005. However, in October 2002 it has been reported that Teledesic would suspend its satellite construction work.

Teledesic aimed to provide an open network for the delivery of the broadband multimedia services of other service providers. It would offer the possibility to other providers to extend their network, both in terms of geographic scope and in the kind of services they can offer. The capacity would be greater than that of the Big and Little LEOs: connections would provide up to 64 Mbps on the downlink and up to 2 Mbps on the uplink, whereas current SPCS offer connection up to 9,6 Kbps.[188]It was going to be an "internet in the Sky". Initially it was projected to launch 840 LEO satellites, however, following this the number of the satellites were reduced to 288 in 1997[189]and has been finally reduced to 30,[190] after the takeover of ICO by Teledesic and the "redesign" of the system. Teledesic would be able to provide services to businesses, schools and individuals everywhere on the planet. Its broadbandness will allow the system to provide 18 simultaneous internet links in each cell, each of which will be 53 Km across,

[185] March 1994. See Robustiano Fernandez, Internet in the Sky, The Ka-Band Quest Continues, in Via Satellite, March 1997.

[186] The Boeing Company invested in 1998 $100 million.

[187] For the stake of Motorola in Teledesic see, The Wall Street Journal, Motorola to take 26% Stake in Teledesic, Friday, May 22, 1998.

[188] See ERO, Mobile Broadband Systems (MBS), 1997, p. 85.

[189] After the merge of Teledesic with Motorola's Celestri. See http://www.ee.surrey.ac.uk /Per sonal?L.Wood/constellations/teledesic.html. Also the site presents analytically the partnerts of the project and investors.

[190] Teledesic has filed for the Ka-Band. The US has licensed 13 systems in the KA-Band. These are Astrolink of Lockheed Martin, Cyberstar of Loral Aerospace, Echostar of Echostar, Eutelsat by Eutelsat Signatories, GE*Star by GE Capital Corp, Intelsat of Intelsat Signatories, M-Star by Motorola, Millennium also by Motorola, Orion by Loral, SES/Astra by SES, Skybridge of Alcatel, Spaceway by GM/Hughes, Teledesic of McCaw, Motorola. Also, Kevin Maney, Where no billionaire has gone before, USA Today, Monday, December 23, 1996.

and 20.000 world-wide.[191] The year 2005 has been now set as the starting year for the provision of services.

| | |
|---|---|
| 1990 | Foundation of Teledesic |
| 1994 | Completion of initial system design |
| 1997 | Allocation of FCC license |
| 1999-2002 | Signing of contract with major manufacturers |
| 2005 | Target year for service provision |

**B. Skybridge**

Another new broadband fixed satellite system for personal communications, which also aimed to make an impressive entrance into the market by providing high speed broadband interactive services to business and residential users, has been Skybridge. Skybridge has been a project of $4bn and 80 LEO satellites[192] elaborated by Loral Space and Communications Ltd. and Alcatel Alsthom SA of France. These two companies joined forces to enter a market in which they were competitors. Skybridge, besides Loral Space Communications and Alcatel, included Toshiba Corp., Sharp. Corp., Boeing and Mitsubishi Corp. of Japan. Its intention was to target small and medium-sized businesses as well as larger business with small, regional sites, at a cost of $US 4,5 bn. The systems would operate in the KU-Band (continuous 1.05 Ghz between 12.75 and 18.1 Ghz for the uplink and continuous 1.05 Ghz between 10.7 and 12,75 for the downlink) and was projected to start operation in the year 2001, year which was however postponed.

Skybridge applied for the FCC filing on February 28th, 1997 and made public its intention to share spectrum with other GSO and FS systems. The system was to provide a wide range of services. They comprise of high speed internet and on line services including information and data bases, teletraining, telelearning and telebanking, video conferencing and video telephony, entertainment (electronic games and video on demand). Furthermore it would be able to provide telecommuting through access to LANs, e-mail and file transfers, LANs interconnection, infrastructure links for telephony, wireless local loop and mobile communications, as well as enhanced narrowband services.[193]

---

191 See The Economist, The final frontier, July 27, 1996.

192 The initial number of satellites was 64.

193 See Skybridge closing in on telecom investors, Satellite International , October 21 1998 In Mid 1998, Skybridge has increased its capacity from 144 Gbps to 200Gbps being able to accommodate up to 20 million users by the end of 2002, when the entire constellation is operational. Sheridan Nye, SKyBridge Doubles Capacity for Broadband Satellite Systems, 1 June 1998, article to be found in Totaltelecom news, www.totaltele.com.

### *VI. Industrialists*

S PCS systems have been operated by the consortia; however, individual industrial entities involved in the projects have also played a very important role, throught strategic moves, acting either as manufacturers, network operators, or as service providers.

#### 1. Manufacturers

Manufacturers have, since the early 90s, managed to become the strongest market players as through supply they have succeeded in penetrating many sectors of the telecom market. Manufacturing has been characterised by a global dynamic, an evolutionary process with the rapid interaction of learning and product innovation.[194]Manufacturers, having better access to technology, could foresee the possibilities of new applications and new combinations of technology and services[195] and as a consequence they could easily penetrate new market segments. [196]Moreover, the fact that they have deep pockets (they can supply infrastructure to many industrialists of the same sectors as well as to others of other sectors), has allowed them to be active in all modern communications branches, as investors, suppliers, manufacturers and customers.[197]

Because of their favourable position manufacturers have been active in building networks and service strategies, making in this way decisive steps towards becoming the main operators of the global networks that have been built for the provision of the services. They have been moving from selling and installing network equipment to the maintenance and service level support and have started building and operating network management centres for new entrants.

The examples in the S PCS industry have been numerous. Lockheed Martin, Loral and Aerospace, have not only been very powerful satellite manufacturers, but they have also been the main investors in various satellite mobile systems, like Astrolink, and Globalstar. Motorola, one of the main players, has been moving in the S PCS industry, in the mobile terrestrial land-based industry as well as in the microelectronics industry. It has been active as the main designer of and investor in Iridium, the second most successful manufacturer and operator for mo-

---

194 Antonelli, Technological change and multinational growth in international telecommunications services, in Noam, Wolfson, Globalism and Localism in Telecommunications, p. 281.

195 See Molony, Vendors poised for global network operations role in Communications Week International, 10/5/99.

196 See Annemarie Roobeek and Jeroen Broeders, Telecommunications global restructuring at full speed, in H.W. de Jong, The structure of the European Industry, Kluwer Academy Publishers, 3rd edition, 1993, p. 273 ff.

197 Stop the world, I want to get off, Public Network Europe, September 1999.

bile telephony and has also been a very active manufacturer of advanced microchips.[198]

**2. Network Operators**

The network operators in the S PCS alliances have been those operating the system's infrastructure, ground and space segment. A distinction could be made between space segment operators (SSO) and satellite network operators (SNO).[199] However, it has been also possible for one network operator to integrate both functions. SSO have been the ones in charge of the satellite or the constellation of satellites and SNO have been the entities (it could be the same as SSO) in charge of operating the satellite network and managing earth stations, control stations and gateways with fixed networks.

This difference may be illustrated in the Iridium case. Iridium LLT was a big company that designed projects, launched satellites and had contracts with a number of national or regional entities.[200]However, at a secondary level was O.tel.O and Telespazio that were in charge of the development of earth stations, type approval, free circulation of terminal equipment services and commercial activities in Europe. O.tel.o and Telespazio could be characterised as SNOs and these entities had to obtain a license to operate in Europe.[201]These SNOs were for Iridium its 12 gateway partners that enabled the system to extend distribution agreements to the local providers in those 12 regions. The gateway providers had their own marketing and sales departments and were the entities that negotiate roaming and distribution agreements with local service providers in each country.

**3. Service providers**

Service providers have been the entities responsible by far for the success of the system. They have been the ones that tried to get the license for the provision of the services. Moreover, they were dealing with the terminal supply, the connection to the local network, and the setting of tariffs, the provision of services, and the customer interface. There have been a big number of local service providers, which gave the distribution a national character. This has been a strategy that the consortia have been following, as it has been easier for the local providers to reach the customers. For instance, about 60 of ICOs investors have been local telecom providers in different countries that collectively hold about 25% of the

[198] Motorola has attracted many clients and is placed behind Nokia and in front of Ericsson.

[199] ETO distinction, in Final Report on Licensing, Copenhagen, February 1998.

[200] Since March 2000 Iridium has stopped providing services and the company has entered into the process of liquidation.

[201] See ETO, Annex 15 to the Final Report on the licensing of Satellite Networks and Services, October 1997, also to be found in www.eto.dk/licensing/r_annex15.htm.

global cellular market. These have mainly been incumbent telecom operators, (Telecom Italy, DT, France Telecom) or mobile cellular operators, like Airtouch, Vodafone.

The European Telecommunications Office (ETO) has divided service providers into two categories. One included entities responsible for the design and the provision of a service. The other included entities that were reselling services and/or transmitting capacities. Service Providers of both categories were not required to obtain any licenses if they were authorised to offer their services through other networks. Service providers' classified resellers would mainly have the same licensing regime as those providing GSM services that is a free regime.

Providers could lease capacity on a wholesale basis to service providers (a service provider could also be the network operator of a gateway) who would then either offer service directly to the users, or further subdivide their capacity for lease or sale to other service providers, who would then provide the service directly to the users. For example, the service provider that held directly the franchise for an entire country might not have chosen to provide telephony and data transmission directly to users, but might have left it to another company to provide position determination services within that country.[202]Similarly, a service provider having the franchise for the whole country might have agreed with a different service provider in each regional subdivision. The final service provider, that is the one accessing directly the customers, would bill them. The elements of the bill could include a fee for access to the space segment capacity, and to the terrestrial network, a service charge for use of satellite airtime and terrestrial facilities and long distance charges.

ICO, for instance, had planned to provide its services to the customers by using networks of national service wholesalers (NSW) or regional services wholesalers (RSW) and national services retailers (NSR). The NSW/RSW would be responsible for organising the ICO distribution to national service retailers and to the end users within a particular country or region. The NSW would be responsible for obtaining the necessary national frequency co-ordination clearances for ICO terminal operation in the 2 Ghz bands and would also be responsible for the installation and the operation of the gateways. The NSR would be the direct interface of the ICO to the end users and would be responsible for marketing and retail sale of the ICO services and relations, including customer credit, billing, accounting and related administration. Globalstar, on the other hand, provided services using the following entities: the gateway operator that provided the

[202] ERO, Mobile Satellite Services, Final Report, Kopenhagen, 1995, p.78.

gateway services, the Globalstar service provider/distributor, that established the distribution agents, the Public Land Mobile Network and Public Switched Telephone Network operators, involved in the delivery of the call, and the local service provider, that sold to the end user. It could have been possible that all the above mentioned functions were performed by one entity.

Service providers were expected to play a key role to the success of the system, as they would be the ones that would have measured the balance of power in the market.[203]They would be the intermediate with the customer, and they would have direct contact with the user. As a consequence, they would be responsible for a big part of market simulation. However, the ones chosen, for instance by Iridium, were inexperienced and did not contributed to a successful marketing and distribution strategy.

[203] Europe, Mobile Satellite Services, Report for the European Commission, March 1998.

# Chapter IV
# Approaching Regulators

## *I. Regulators - Strategic actors that deal with law*

Regulators perform a complicated task. They not only have to examine in what way rules can be best used, they also have to take into consideration all factors that affect the elaboration of the rules, that is interest groups, organisational structures, decision making processes, as well as assessing the further impact of regulation on societal structures.

With the general trend towards lighter regulation, the increased competition has brought about the need for effective and independent regulators, which act as strategic planners that define future market structures particularly focusing on the following topics:

- efficient management of shared resources
- unrestricted market opening
- a juridical regime to be applied in all relationships among operators and users
- the independence of political power and commercial interests in order to guarantee a healthy market

The formation of a S PCS regulatory framework has been a challenge for the regulators given the complex technology, the number of actors involved and the national divergences in terms of frequency allocation and assignment methods, licensing conditions, fees and custom duties.

The main regulatory bodies that have played a significant role with regards to S PCS and will be examined in this chapter will be the following. First, ITU, the sole S PCS global regulatory body. Second, at regional (European) level, the European Commission and CEPT with its two committees, the European Radio-communications Committee (ERC) and the European Conference for Postal and Telecommunications Authorities (ECTRA).[204]

The purpose of this chapter is to present the various regulatory bodies, describe their functions and analyse the role they have played with regards to the introduction of the S PC systems at the various policy making and regulatory levels. It will be made clear that the introduction of S PCS systems in the Community has increased the quantity of communication paths among Member States, European Community institutions and industrialists, who have been the players mainly involved. This increase was contributed to by the creation of committees of expertise (i.e., Licensing Committee, Milestone Review Committee) and the multiplication of co-operation and negotiation procedures. These helped the Commis-

---

[204] These are European and not European Community Institutions.

sion and the ERC to take regulatory decisions and to guarantee the participation of national experts in decision-making procedures. Moreover, there have been regular meetings and consultations between the EC, ERC and national administrations with not only European, but also US and Japanese industry.[205]

**1. Global Regulator - International Telecommunications Union**

As Karl Deutsch proposed almost forty years ago, telecommunications infrastructure constitutes the nervous system of modern society, over which various groups have gained control, as new centres of political, economical and social power are to be found among those that control communications means.[206]The history of the International Telecommunications Union, which has always been the main international regulator, has been the history of satellite policy-making. This has been a history of struggle for control over frequencies and orbits, since for governments control over frequencies has meant increased political power; a nervous system with many fragile counterbalancing forces.

The ITU has been the oldest intergovernmental organisation,[207]a forum where for almost a century the capacity of international policy making has been tested. Its history is related to some important issues that have troubled the world community regarding the management of scarce resources, such as spectrum and the geostationary orbit.[208]The ITU was founded in 1865, after states realised that is-

[205] In particular, during these interactions, the same actors have participated in different law-making procedures. For instance, the same industrialists met the European Commission and lobbied with delegations of developing countries (Teledesic, Alcatel, Ericsson); the same national delegations participated in the licensing committee or other meetings, and lobbied with governments and industries during the ITU conferences.

[206] Karl Deutsch, The Nerves of Government, New York, 1963.

[207] The International Telecommunications Union was established in 1932, after a merger between the International Telegraph Union and the International Radiotelegraph Union. See Lyall Francis, The role of the International Telecommunications Union, in Lafferanderie, Outlook on space law in the next thirty years, W. Emery, National and International Systems of Broadcasting, their history, operation and control, 516-531, 1969, Pierre Jaccard, Les bases et les sources du régime juridique des télécommunications, Diss., 1991. For an analysis of the structure of the ITU, see also James Savage, The Politics of International Telecommunications Regulation, Westview, 1989.

[208] As regards the spectrum management the question has always been how to distribute the allocated frequencies. This problem became obvious after 1976, after the so-called "Bogota Declaration", which signaled the beginning of the North-South conflict, the struggle between the developed and the developing countries. The reason for this conflict was ITU itself and the procedure it had established for the frequency distribution, the well known "first-come, first served" principle. According to this principle, the frequency was allotted to the first country, which expressed interest and made a claim. The first countries were the developed ones, as they possessed the necessary technology, which would enable them to exploit the frequency. On the contrary, the developing countries did not. As the frequencies were allotted

sues related to communications needed international co-operation. It was officially recognised as an UN-international institution during the Atlantic Conference in 1947.

The ITU constituted an intergovernmental body of complicated structures, multilateral in character. Its function could be looked at from various perspectives. First, as a lato sensu legal system. The rules and regulations have been voted by the 189 member states and have been applicable to the ones that commit themselves to it.[209] Second, as a negotiating system which has been trying to bring together actors with different interests and promotes the common interests of states. Third, as a construction of high political impact: common or conflictual interests has brought states together to regional clusters and showed how the developed countries could influence the decisions on policy and regulatory issues of the developing ones.

The ITU was born in an era of absolute national monopoly. The National Telecommunications Organisations (NTOs) had a monopoly status in their domestic markets and this forced foreign carriers to terminate the traffic they had carried

---

to the developed countries, eight developing ones (the equatorial ones: Brazil, Zaire, Uganda, Congo, Colombia, Kenya, Indonesia) reacted by signing the "Bogota Declaration". In this declaration they raised claims of "property-isation" of the geostationary orbit and at the same time posed serious questions about the international space law. They claimed that the geostationary orbit belongs to their state sovereignty, so all other states that wanted to launch a satellite in the geostationary orbit should first ask their permission. Furthermore they also raised claims of a technical nature. They characterised the geostationary orbit as a special orbit, caused only by the phenomenon of gravity, an argument which proved to be wrong, as years of studies showed that the orbit is caused by various natural forces. The event challenged ITU and international space law, as the question was posed whether the geostationary orbit was part of space, which meant that it was submitted to the regime which governs space, or if it formed part of state territory, which meant that it was an administrative responsibility. Applying space law would mean applying principles concerning space. These principles, according to the most important legal text concerning space, The Outer Space Treaty (1967), would be that it is the province of all mankind (Article I) and that it must be used on the basis of equality and of international cooperation (Articles IX, X, XI) in accordance with international law (Article III) and in the interest of all mankind (Articles. I, II). ITU reacted in the next conference (in 1977) by distributing orbit position and frequencies to Regions 1 and 3 (Europe, Africa, Australia, Asia) and then five years later, during the Nairobi Conference in 1982 to Region 2 (America). See also Fawcett, Law of Outer Space, 1984, Dahinden, 1990, Zhukov Gennady and Yuri Kolosov, International Space Law, New York, 1989.

[209] There are also 600 Sector Members. These are organisations with interests in communications, that is, major players including manufacturers, service providers, network operators, infrastructure designers representing the relevant Members. For instance from Germany Alcatel, DT, E-Plus, Manesmann Mobilfunk, Siemens, from Finland Nokia, Finnet, Sonera, Tellabs Oy and many other examples.

to the national frontier.[210] The idea of the global market did not exist. The ITU did not develop the idea of transnational communications, or reduce sovereign rights. It only tried to provide a solution to the technical problems of interconnection. Consequently, it was necessary to develop common technical standards and to reach international co-operation.[211]

Among its tasks the ITU had first, to promote international co-operation in the telecommunications sector, and regulate the most efficient use of technical facilities. Second, to effect the allocation of the radio frequency spectrum, and the registration of the radio frequency assignment, in order to avoid harmful interference between the radio stations of different countries. Third, to contribute to the adoption of measures in telecommunications so as to ensure safety in the relevant services and fourth, to elaborate, study, adopt resolutions and recommendations and to collect public information concerning telecommunications matters.[212]

The ITU provided a forum where all interested actors, including states, industries, international organisations could meet and look for partners to conclude further agreements.[213] However, interconnection and transnational communications were regulated through bilateral agreements among states, and not yet at international level. National administrations regulated the domestic and international issues of interconnection of their country, and represented it to the ITU. However, in recent years this situation started to change. Technological progress, liberalisation and deregulation have had a strong impact on ITU, national telecommunications organisations have lost their dominance and powerful industrial players have made their appearance.

The ITU has been characterised by the following features. First, there has been a very large number of participants. With 189 members, plus the 600 sector mem-

---

[210] See Oliver Stehmann, Network Competition and European Communications, Diss., EUI, Florence, 1993, p.282.

[211] On the history of ITU, see International Telecommunications Union, ITU: A hundred years of cooperation, Geneva, 1965, also ITU, The new world satellite order: a documentary review, with commentary of the 1985 World Administrative Radio Conference, 1986, The Procedural Aspects of International Law Institute Washington, D.C, p. 6 ff.

[212] During its long existence, ITU has elaborated the so called Radio Regulations. It is a long text of rules and regulations, concerning the technical and regulatory aspects of telecommunications, consisting of more than 1800 pages, These regulations have been elaborated during the World Radiocommunications Conferences. Issues relating to frequencies are registered by the International Frequency Registration Board, a body of ITU.

[213] Bilateral agreements was the way states used to deal with technical issues, such as standardisation and interconnection. The ITU developed the technical standards and the states negotiated bilateral agreements with other states.

bers and the neutral observers, the ITU has constituted one of the biggest international organisations at a world wide level. Second, the points of conflict have raised many important issues, needing a long time to be resolved. One of the most common points of conflict has been the debate between the developed and the developing countries, or the North-South conflict, as it is known.[214] Third, the complexity of the procedures, (unavoidable to a certain point due to the large number of participants), the bureaucracy, the hierarchical structure, and the increased number of committees within the ITU, have been additional obstacles. Fourth, in the last years there has been a rise of strong regional coalitions. Common interests stemming from common geographical regions have resulted in countries from the same regions forming alliances by claiming the same frequency rights.[215] As a result, regional interests have been strongly supported by various important regional blocks: CEPT, APT, CITEL, Arab league.[216] Fifth, the ITU, because of its flexible nature, the big number of actors and its strong political interests, has been lacking of a dispute solving mechanism that would make the organisation an arbitrator with a global status.[217] One of the reasons for this has been the system for frequency allocation, the "first-come, first served", according to which the first country that made a claim was best served, since the frequency was allotted, registered and protected under its name. This method became an obstacle to the international attempt to introduce competitiveness into the spectrum, since it introduced monopolistic privileges for using a frequency, which had sometimes nothing to do with the real needs of the countries.

As far as the S PCS regulation is concerned, the involvement of ITU has started since 1992,[218] where important policy and regulatory decisions have been taken.[219] The ITU, especially during the last decade, where economical forces

---

[214] See above, fn. 211.

[215] One of the reasons for these formations has been the voting system of the ITU, where one country equals one vote: the more the countries in the coalitions, the more the votes. See Liching Sung, WARC-92: setting the agenda for the future, Telecommunications Policy, November 1992, p. 624 ff.

[216] On the ITU's voting system one country one vote, see Lyall, The Role of the International Telecommunications Union, in Lafferanderie, as above.

[217] However, this flexibility may have positive effects on the parties in conflict and lead to a faster and more efficient way of solving the problem, like in the case of the dispute among Iridium and the Astronomers, where the parties ended up solving their conflict by signing a "time sharing agreement".
See Scherer, Innovationsoffenheit der europäischen Telekommunikations Regulierung-Konzepte und Instrumente am Beispiel der S-PCS-Systemen, in Wolfgang Hoffmann-Riem (ed.), Innovation und Telekommunikation, Hans-Bredow Institut für Medienforschung, Hamburg, 2000.

[218] With the WARC 1992 in Spain.

[219] See below, Chapter IV.

dealing with S PCS made their appearance, has revealed a new world power, based on the polycentrism of power. Regulatory institutions have been playing, together with the industrial sector, a multilevel game.[220]The role of private actors has been strengthened and public actors have realised that very powerful private actors have entered the playground. Co-operation among actors intensified through numerous meetings and conferences, which had the objective of speeding up the law-making processes.

**A. ITU Regulation and Procedures**

The ITU regulation has been covering all issues of satellite communication, and in particular, radiowaves, frequency allocation for radionavigation, broadcasting, S PCS communications and orbit assignments.

a. Frequencies

Since 1947 the ITU has been regulating frequencies and orbit positions.[221]Frequency regulations have constituted one of the main contexts of the RRs that have been dealing with all issues concerning the emission and reception of radio waves, frequency and orbit allocation.[222]These provisions regulate issues such as international protection from harmful interference of the recorded frequencies (WARC 47), frequency allocations for space research purposes (WARC 59), prenotification procedures for terrestrial and space services sharing the same bands ( WARC 63), the establishment of equal rights of all Member States to frequency access for fixed satellite services (WARC 71),[223]regulation of broadcasting services (WARC 77),[224]regulation of mobile satellite service (WARC/

[220] See Liching Sung, WARC-92: setting the agenda for the future, Telecommunications Policy November 1992, p.624 ff.

[221] For instance, frequencies were assigned for the first time during the WARC 1959 for various space activities and research purposes. In WARC 63 the prenotification procedure for the use of terrestrial and space activities was elaborated. During the WARC 1971, the term "broadcasting satellite services“ was defined and frequencies were also allocated for these services. Furthermore the issue of the „spill-over“ effect was also discussed and frequencies were allocated to space radio communications. In WARC 77 frequencies were allocated for direct broadcasting satellites (11,7 Ghz-12,5 Ghz). During the WARC 1979 the Radio Regulations were revised and new frequencies were allocated to fixed satellite services. WARC 85/88 worked on satellite mobile services.

[222] They include provisions for maritime mobile services, radio astronomy, land mobile services, nautical mobile services, radio navigation services, space research services, meteorological services and all other kinds of satellite services. See Steward White, Stephen Bate and Timothy Johnson, Satellite Communications in Europe: Law and Regulation, 1996, ch.3.

[223] In regards to the debates on FSS, they were finalised during WARC ORB 85/88.

[224] The plan for the allocation of frequencies for broadcasting purposes was drawn for regions one (Europe, Africa) and three (Asia, Oceania). The band allocated was the Ka-Band. For every state, five satellite frequencies were distributed. The frequency band was divided into four channels with every 1 MHz a bandwidth. Every satellite orbit position was therefore to

ORB 85/88), S PCS frequency allocation, Resolution 46[225]( WARC-92), Giga LEO frequency allocation (WRC-95), "administrative due diligence procedure" applicable to S PCS (WRC-97).[226]

Although the ITU has been trying to operate as a neutral independent mediator, decision-making procedures have always been influenced by interest groups, which in the previous decades, were mainly made up of national governments. This made the whole procedure vulnerable, since it was the governments that determined the power distribution of the ITU, showing to what degree states can generate regulatory changes.[227]However, in the 1990s the global order started changing, because of the relocation of power among national governments and industrialists. This transformation was clearly reflected in the ITU conferences, where backstage political games and economic interests determined the rules of the game for the S PCS, which had emerged at the beginning of the nineties.

**WARC 92**

The WARC 92 was a landmark conference which signalled a new era in international spectrum policy and in satellite communications in general. Regarding the conference itself from a technical and procedural point of view, it was the first WARC in more than 12 years to attempt a broad revision of the international table of frequency allocation. Furthermore, as opposed to the WARCs of the 1980s, which generally concentrated on services, WARC 92 addressed a wide range of allocation issues covering many segments of the radio spectrum. Moreover, this WARC allocated frequencies to LEOs. [228]

---

be used with 40 channels. This meant that at maximum eight states, each one with five channels, could commonly use an orbit position. Frequency allocation for Region 2 (America) took place in 1982 at the Nairobi Conference.

[225] A decision which left the right to decide if they want to participate in the procedures for the NGSOS up to the Member States. European Commission, DG XIII, Satellite Personal Communications and their consequences for European Telecommunications, Trade and industry, 1994. p.202.

[226] This resolution required administrations to submit „administrative due diligence information to the ITU Radiocommunications Bureau, related to networks and services. Information should be provided on the space craft manufacturers, the delivery window for the spacecraft, the launch date and other. The form asked for the precise date that frequencies would be brought into use. With this requirement the ITU expected to decrease the number of „paper satellites." However, this regulation applied only for geostationary satellites. See Gerald E. Oberst Jr., New Forms, New Fees, Via Satellite, September 1998, p. 18.

[227] See John Savage, The politics of international telecommunications regulation, 1989, Georgia Chioni, Regulating Frequencies, Esperia Publications, 1999.

[228] WARC 92 considered little LEOs and Big LEOs. Little LEOs planned to provide data and position location services and would operate on the VHF/UHF frequencies. Big LEOs would provide telephone services in addition to data. In fact, besides the allocation that was made for MSS, the following frequencies where granted to the big LEOs: 1610-1626.5MHz (down-

The important issues that marked the conference were the following. It was the first conference where frequency bands were assigned to telecommunication satellites not orbiting the geostationary orbit. The insistence of many delegations on regulating bands for the LEO satellites, which are used for the needs of mobile telephony, made mobile services the distinct winner. Fixed satellite services were reduced to lower frequency bands or had to convert to optical fibre.

Second, the existence of strong political influence and private commercial interests was very obvious.[229]Companies with big investment and big interest in satellite mobile communications used various tactics to pursue their aims, such as enlisting the support of their own governments, who then engaged in diplomatic fights to promote their own interests (Motorola, Ericsson, and Alcatel).[230]Furthermore, they tried to ensure the participation of key personnel from the industrial sector as official delegates of national governments and even sent fleets of lobbyists to the conference to persuade delegates from less developed countries to support their projects.

During the conferences, countries realised that being in the same region meant having common interests. As a consequence, they started forming regional groups, blocks and coalitions in order to represent their interests better. For example, EC interests are represented through CEPT and the European Community is only an observer. Other important groups have been the Arab league, CITEL, and the Asian and Pacific States. In this context, governments and private actors have been particularly concerned about the rise of strong, regional blocks of countries that are uniting to protect and promote their economic interests in international meetings. As in the ITU, it is one country one vote, and the creation of large coalitions can easily affect decisions. It was also the first time that the CEPT block[231] of European countries spoke with one voice, defending common European interests and succeeded in getting what it wanted. This was the first sign of the formation of a strong European policy, which since then dominated in the WRC conferences.

WARC 92 revealed the power of the users, which meant that markets would be demand-driven and pointed out the existence of powerful actors-users. This is-

---

link), 2438.5-2500 Mhz(downlink) 1613.8-1626.5 MHz and for little LEOs, 137-138 Mhz (downlink), 148-149.9 Mhz (uplink) and 400.15-401 Mhz (downlink). See, ITU, WARC-92, Final Acts.

[229] See U.S. Congress, Office of technology assessment, The World Administrative Radio Conference: Technology and policy implications, Washington, 1993, Chapter 1.

[230] See Liching Sung, WARC-92: setting the agenda for the future, Telecommunications Policy, November, 1992, p. 624 ff.

[231] Analitically on CEPT below in this chapter, also Liching Sung, as above.

sue has created important changes in the role of the ITU as the international spectrum regulator. It became obvious that negotiations which had previously only involved governments would, in the future, include a variety of spectrum special interest groups (private sector and extra-governmental organisations that represent important international users of radiocommunications) more directly. The message that came from WARC 92 was how difficult it had become to achieve international consensus on spectrum matters. Moreover, the most important issue in WARC 92 was the battle between the existing and the new technologies, as the existing technologies did not want to be left aside because of the new ones and at the same time, the new ones wanted to be accommodated.

**WRC 95 - WRC 97**

In WRC 95 the new division of powers, which had started with WARC 92, continued and some very important decisions affecting MSS services were made. The most important of them was the allocation of 20Mhz, which was brought forward to the 2Ghz band for the MSS.[232]

The developing countries resisted at first, but the US intervention, assisted by some other developed countries, managed to prevail. Furthermore, spectrum was allocated to the broadband MSS, for which Teledesic was the only candidate, who, with the support of the US government, was probably the strongest lobby before and during the conference.[233]

The main results of the WRC 97 were the following. First, there would be competition in the global provision of satellite broadband communications and second frequency requirements were included for the Universal Mobile Telecommunications Systems (UMTS) on the agenda of the forthcoming conference.[234]

---

[232] See ITU, WRC-95, Final Acts.

[233] On this issue the CEPT had no position. From an economic and political point of view, during the conference, it was obvious that the US delegation was determined to fight for its non geostationary satellite systems, and its advantage was that its demands were backed by the majority of the developing countries. As there was no European interest for such a system, the CEPT was not very interested. In particular, the Commission was in favour of producing a document in order to have a position on the issue of Teledesic during the WARC. The CEPT, however, wanted to achieve the following points: first, it wanted to achieve a freezing time, which meant that it wanted to allocate the frequencies, but discuss the issue during the next WRC. The second, objective of CEPT was to avoid voting for Teledesic, and the third was to prevent a US spectrum monopoly. As a consequence of the difference of opinion between the EC and the CEPT, they both claimed that the US delegation had brought the issue of spectrum distribution very late and that the EC was not in a position to co-ordinate a common position and that the interests were not adequately secured.

[234] On the characteristics ans the importance of UMTS as third generation of communications, see Decision No 128/199/EC of the European Parlia;ent and of the Council of 14 December

This was in line with both Europe's agenda and policy for the further development of the mobile terrestrial communications. From the above-mentioned observations, the WRC made it clear that policy-making would focus on the two most important issues of satellite communications, namely the satellite broadband communications systems and the International Mobile Telecommunications (IMT)2000/UMTS.[235]

Second, the interplay of commercial interests increased at this conference, not only in terms of marketing, and lobbying, but also in terms of the participation of the industry in national delegations.[236] In this conference, Europe did not have to confront the US, as both the US and Europe shared the same expectations, which were to get frequencies for broadband satellite services, which were to be provided by US and European projects. As far as the CEPT was concerned, it was the first time that CEPT was very well prepared and all 43 Members had put their signature to the common proposals.[237] Moreover, CEPT argued that the satellite issues should not be given exclusive priority, as the US wanted, but that power limits should be established, so that other services were included in these bands.[238]

**2. Regional Level - Europe**

In Europe, there have been two networks of policy making. On the one hand the European Community, with the Commission being the leading actor in shaping policies and on the other hand CEPT through ERC and ECTRA assisted by the Licensing Committee and the MRC, two committees of expertise, developed in different institutional frameworks. These regulatory and administrative institutions have been closely assisted by the industry, which through a working group

---

1998 on the introduction of a third-generation mobile and wireless communications systems (UMTS) in the Community, L. No 17/1, 22/1/99.

[235] See ITU, WRC-97, Final Acts. What the last WRC conferences have demonstrated has been the fact that political decisions could easily be manipulated. Moreover, they have shown that industrial lobbying cannot be stopped, as industrialists have become the decisive backstage players and the more powerful resource an enterprise is, the bigger its influence on political actors. See, for instance, the Tonga case, when Tonga filed for the remaining sixteen orbital slots that link Asia with the United States, and Intelsat reacted. See Henry Wang, The Paper Satellite Problem, Journal of Air Law and Commerce, 1998, v.63, p.853. See also Vineeta Shetty for Communications international, ITU Plenipot 98 plays Paper satellite Ping-Pang, 29 October 1998.

[236] For example the major satellite broadband consortia, Skybridge and Teledesic, had lobbied heavily around the world in preparation for this conference.

[237] There were also disputes of a private nature, such as, the dispute between Astra and Teledesic, and Astra with Eutelsat for frequency bands.

[238] However, it was finally agreed that power limits should be established, but not for the Teledesic band.

has tried to help the regulatory institutions. Regulatory efforts at European level have been assisted at national level with the help and cooperation of national regulatory authorities, as well as the responsible in each country ministry.

**A. European Community**

a. Theoretical framework

The emergence of the S PCS regulation has been triggered by multilevel governance transformations at an EC level. These have signalled a re-location of political, regulatory and economic power. The reasons behind this have mainly been the following. First, there has been a change at institutional level from a supranational base, as was the case for satellite communications, to a structure where the intergovernmental element appeared to be stronger, and influenced by functional and institutional specialisation.[239] Second, there has been a multiplication of the procedural and regulatory links between the various actors, caused mainly by the existence of many interest groups, leading to a new definition of the role of public and private actors, and to a redistribution of competencies. Third, there has been a shift from the "rule of the leader", to the increased importance of negotiations and bargaining.[240] Fourth, there has been a change in the role of regulators, from a dependent institution to a more independent body.[241]

In regards to satellite (S PCS) communications, paths of conflict and co-operation best illustrate the EU as a system of multilevel governance, which may be indicated as follows:[242]

- national and supranational actors. The division of powers and the distribution of competencies lead to implementation problems. The delay in implementation implied that issues of national sovereignty had arisen, and were related to important sectors for the national economies, such as frequency assignment, licensing fees and custom duties imposed on mobile earth stations entering a country. In the last years, however, Member States have at large implemented the Community's measures within the time-frames set, as they have realised that harmonised procedural rules help the formation of a strong European

[239] through the emergence of many committees of expertise that have assisted the European Institutions in specialised issues related to telecommunications. The Milestone Review Committee (MRC), the Licensing Committee, and the Open Network Provision Committee have been such examples.

[240] Mainly through unofficial meetings between regulators and economic actors, such as the European Commission and S PCS industrialists.

[241] For instance, the ERC, which has emerged as the soft law regulator responsible for harmonising S PCS and UMTS policies within the Member States.

[242] See Gerhard Fuchs, The European Commission as a corporate actor? European Telecommunications Commission after Maastricht, in Mayes D. (ed.), The evolution of rules for a Single European Market, 1995, pp. 171 ff.

(satellite) industry.[243] Such an example has constituted the abolishment of fees on the circulation of the mobile earth stations, pan-European harmonised procedures, simplified type-approval procedures.

- The Commission had been warming up relations with national representations and units, as in the case of the NRAs and the Licensing Committee. The NRAs have become the important national regulatory bodies, responsible for licensing and spectrum assignment. The Licensing Committee, comprised of national representatives, mainly from the NRAs and with the participation of Commission and CEPT representatives, has taken important decisions on S PCS. For instance, the Licensing Committee approved of the mandates on S PCS (on the introduction of Big LEOs, Little LEOs and Mega LEOs).
- There has been a considerable increase of Committees of expertise, composed either of national representative, or by industrialists or various experts, who mediate and assist the Commission and national regulatory units, such as the Milestone Review Committee, the Open Network Committee, the SIG OSS Sat, SIG Mr Sat and others.[244] The role of the MRC is particularly important, as it supervises the application procedures of new S PCS operators in the European Community.
- There has been a significant increase in interaction between regulatory bodies and economic actors. Industrialists (satellite manufacturers, space segment operators, S PCS operators and providers) participate in many meetings of policy and in regulatory bodies and assist the Commission in identifying regulatory barriers in the various Member States that hinder the satellite (S PCS) communications.[245]

b. European Commission

Until the mid 1980s the telecoms were considered to be a public service, where the Treaty's competition rules were not applicable.[246] Since the beginning of the 1980s, however, the European Commission has been developing a wide range of activities in the telecommunications sector, pointing out problems and establishing an agenda for the measures that needed to be taken. This strong involvement has resulted in the development of a dynamic policy and regulation, for all fields of telecommunications, such as infrastructure, mobile communica-

[243] See, for instance, the R&TTE Directive and COM(99)537.

[244] Both the SIG OSS Sat and the PT-GAIL have been committees of expertise created in the framework of ECTRA activities. The SIG OSS Sat dealt with satellites and One-Stop-Shopping procedures, whereas the PT-GAIL has been a project team on general authorisations and individual licenses.

[245] See the report of the SAPRWG on Market barriers.

[246] Natalicchi G., Telecommunications policy and integration processes in the European Union, New York, Diss., 1996, p. 367.

tions, personal communications, voice telephony, advanced services and technological equipment, UMTS.[247]

The Court of Justice (ECJ) assisted this effort. Since the mid 1980s the Court has challenged the notion that telecoms are a reserved service sector, has extended Treaty competition principles in the sector, and affirmed the Commission's power to reinforce them. In particular, through its decisions, it has reaffirmed the Commission's prerogative to legislate in the telecommunications sector. [248]

The Commission initiated a liberalisation process based on -according to many analysts- an ambivalent competence.[249]The release of the "liberalisation directi-

[247] Mark Thatcher, The interaction between the EC and its Member States in telecommunications regulation, paper presented in the workshop "The Aftermath of Liberalization: Multilevel Governance and the Regulation of Telecommunications", Florence, 14 November 1997, Sauter Wolf (1993), The telecommunications law of the European Union, European Law Journal 1, 1, pp.92-111, Thatcher Mark, The Development of European Regulatory Frameworks: The expansion of European Community Policy-Making in Telecommunications, in S. Stauridis, E.Mossialos., R.Morgan and H.Machin (eds.), New Challenges to the European Union: Policies and Policy-Making, 1997.

[248] In the British Telecom case, where the Court upheld the Commission's decision to stop BT from pursuing monopoly practice in the UK. See Case 41/83, Decision of March 20, 1985, based on the decision of British Telecoms of 1982, OJ 82/861. In 1988, France, Italy, Belgium, Germany and Greece introduced a complaint with regard to the Commission's Terminal Equipment Directive which liberalized the terminal Equipment markets. The court upheld most of the parts of the directive and reaffirmed the right of the Commission to order states to remove the exclusive rights granted to their NTOs. In 1991 the ECJ released its decision concerning the Terminal Equipment case (French Republic and others v. Commission of European Communities [c-202/88, 1992, 5, CMLR 552). France, supported by Belgium, Germany and Greece, applied for an annulment of certain provisions in the Terminal Equipment Directive (88/301/EEC of 16/5/88 on Competition in the Market in Telecommunications Terminal Equipment, OJ 2. 5. 95 L131). The ECJ confirmed that (ex. Art. 90(3)) Art. 86.3 of the Treaty empowered the Commission to specify general obligations with which Member States must comply under the Treaty by the implementation of directives. The second case has been that of the complaint of Spain, Belgium, Italy and France against the service directive. The court ruled in favor of the Commission, reaffirming its right to enforce such specific legislation. See joint cases: C-271/90, 281/90, 289/90, Spain,Belgium and Italy vs. Commission. Court decision 92/ 326/08. For more details see Natalicchi, p. 368-9. For analysis of this issue see, Joseph Fesenmair, Öffentliche Dienstleistungsmonopole im europäischen Recht, Berlin Verlag, 1995, Naftel James Mark, The natural death of a natural monopoly: Competition in EC after the Telecommunications Terminal Equipment Judgment, ECLR 1993, S. 105, Piet Jan Slot, Case Studies: C-202/88, C-41/90, C-260/89 Common Market Law Review, 28, 1991, 964-988.

[249] The Commission was accused of being an "undemocratic core that wants to run a federal group of states". See Vineeta Shetty, Cover Story, On a different Wavelength, Communications International, 1.5.99, p. 6.

ves" by the Commission, were based on Article 86.3 of the Treaty, a mechanism which allowed the Commission to take legislative initiatives by bypassing the Council. Initially the policy of using Article 86.3 was applied through the Financial Transparency Directive, which dealt with possible subsidies of the Member States' in public undertakings.[250]Since 1988[251] the Commission has been using Article 86.3 for the liberalisation of the telecommunications market.[252]By using this article, the Commission wanted to counterbalance the Community's competition policy and the Member States' sovereign economic rights, and to attack the exclusive rights of the NTOs. In this way, the Commission would establish a successful supranationalism, which would lead to the "Europeanisation" of a formerly large nationally shaped sector.

In particular, the Commission wanted to ensure that Member States neither enacted nor kept in force any measures, which were contrary to the Treaty rules, respecting NTOs as exclusive right holders. It was actually an attempt to force the Council to liberalise the telecommunications market and regulation. This did not occur without a reaction from Member States. As has been pointed out, the Commission, by using Article 86.3 and the competition articles in general, became in this way a «powerful regulator».[253]As a consequence, important directives, those that formed the basis of the liberalisation in the telecommunications fields, were based on this article: the Services Directive, the Terminal Equipment Directive, the Satellite Directive, the Full Competition Directive.[254]

The Commission has tried to build a coherent European approach to the telecommunications field, not only with legislative initiatives such as the Green Paper on telecommunications, but also with ambitious programs such as

---

[250] European Commission, Financial Transparency Directive of 25 June 1980,(80/723/EEC), confirmed through the ECJ, Dec. 188-190/80, 6.7.1982, B. 1982, 2545 and ECJ, Dec. 118/85, 16.6.1987, B. 1987, 2599, also broadened through the Directive of the Commission of 24. July 1985 on amendment of Directive 80/723 on the transparency of the financial relations between the Member States and public undertakings, 1985, No L 229/90 and the Directive of the Commission of 30 September 1993 amending the directive 80/723 on the transparency of the financial relations between the member States and the public undertakings , 1993, OJ L 254/16. On the issue detailed see Calliess Christian, Ruffert Matthias (eds.), Kommentar des Vertrages ueber die europaeische Union und des Vertrages zur Gruendung der Europaeische Gemeinschaft, Art. 86, Darmstadt, 1999.

[251] See the Terminal Equipment Directive, 88/301/EEC.

[252] See Calliess/Ruffert, (ed.)., Article 86.3.

[253] See P. Holmes et al., International competition policy, Telecommunications Policy, vol.20, No. 10, 1996, p.759.

[254] Services Directive, 90/388/EEC, OJ L No 192/10, Terminal Equipment Directive, 88/301/EEC, OJ L No 131/73, Satellite Directive, 94/46/EC, OJ No L 268/15, Full Competition in the field of mobile and personal communications, 96/2/EC, OJ L 20/59, 26.01.96.

EUREKA, RACE and ESPRIT.[255]Furthermore the Commission has realised that a market-based economy would be generally a more efficient way of introducing competition and would lead to reduced prices for the consumer.[256]This market-driven strategy was a compromise between opposing governing regimes and two powerful groups. On the one hand the demands of the public operators were met, as they were keeping their privileges in terms of infrastructure provision, while on the other, the demands of the large business users were respected by gradually introducing competition for services and equipment into the market.[257]

In regards to the introduction of S PCS in the European Community, the Commission has been the leading actor in introducing the new satellite systems. Since the early nineties it made the necessity for a legal framework obvious,[258] pointed out the need for co-operation among the interested actors, and emphasised on the importance of supervision of the works of the CEPT.

c. The Licensing Committee

The Licensing Committee was created through the provisions of the „Licensing Directive" and had a legal personality.[259]The Licensing Committee was not a policy maker like the European Commission, it co-decided. It was formed and functioned under the auspices of the European Commission. The importance of the Licensing Committee lies in the fact that Member States participated in regulatory initiatives.

It has been composed of national delegations (administrators or experts)of the member states. The licensing committee had a decisive role. As it was in close co-operation with the Commission, it decided on all important issues on regulation, taking the needs of national governments and industries into consideration on important issues related to S PCS, like the mandates of the Commission to the ERC.[260]Moreover, it brought together Members of the Commission, CEPT representatives and the Member States, which had to approve or express their doubts about all proposed measures.

---

[255] Bruce R. Robert, Cunard P. Jeffrey and Director D. Mark, From telecommunications to electronic services, a global spectrum of definitions, boundary lines and structures, 1986, Butterworth, p. 125.

[256] Stephan Bate, Stewart White, Timothy Johnson, Satellite Communications in Europe, 1996.

[257] Hulsink, Do nations matter in a globalised industry?, 1996.

[258] The introductory document has been a communication of the Commission of 1993 on the introduction of the S PCS in the Community [COM(93)171]See below, chapter IV.

[259] Articles 14, 15, 16 of the Licensing Directive, 97/13/EC, OJ L No 117/15.

[260] The mandates have been based on a mechanism established through the S PCS decision. See below chapter IV.

## B. Europe: CEPT - The role of CEPT and its subcommittees - Intergovernmental Regulators at European Level

a. CEPT

CEPT was involved in the regulation of telecommunications, as the exclusive forum of co-operation, long before the beginning of the Community's policy activities in the field.[261] It was created in 1959, on a common consensus of Member State administrations of Posts and Telecommunications, to deal with technical issues related to telecommunications. Despite the great diversity that has characterised the national telecommunications organisations in Europe, the variety of structures, the nationalism of the European industries and the national administrations, the European States have succeeded in establishing very good relations amongst themselves. They have managed to co-operate, and as a consequence, the European Conference for Post and Telecommunications was created to deal with the exclusive domain of the traditional Post, Telephone and Telegraph Administrations.[262]

The tasks of CEPT were mainly technical, as it had to deal with the setting of tariffs and other relevant issues among the NTOs.[263] However, it is important to point out that CEPT has not been like the ITU or the European Community, a „treaty-based organisation". The basis of the CEPT arrangement has been that of an administrative agreement between organisations, the decisions of which the national regulatory authorities of its members states are not obliged to implement.[264]

In the past fifteen years CEPT has been reorganised: Prior to 1987 CEPT was composed of telecommunication regulators, systems operators and telecommu-

---

[261] See also Dang-Nguyen, Schneider and Werle, Corporate Actor Networks in Telecommunications, Max-Planc Institute für Gesellschaftsordnung, Köln, 93/4, p.13.

[262] The protocol was signed in Montreux by nineteen countries. It was agreed in the arrangement that "lorsque les administrations d'un pays sont distinctes, elles sont considérées comme une seule administration membre".

[263] Long distance calls, common positions in international fora, technical standards or interconnection between networks. Actually the objectives are defined in Article 4 of the arrangement: "1. La conference a pour objectives essentiels le resserement des relations entre les administrations membres ainsi que l'harmonisation et l'amelioration pratique de leurs services administratifs et techniques, 2. Entrent notamment dans ses attributions: la réalisation de travaux et de services d'intérêt commun; la simplification et l'amelioration des services postaux et des télécommunications; la mise à l'etude des questions concernant l'organisation, la technique et le fonctionnement des services; l'examen en commun des propositions présentés aux congrès et conférences des organismes internationaux des Postes et Télécommunications; l'echange d'informations et de fonctionnaires." See also L.J.Libois, Genése et croissace des Télé cmmunications, Paris, Mason, 1983, p.294. On CEPT, see http://www.cept.org.

[264] ERO, Spectrum below 2,5 GHZ, Report, October 1997.

nications services providers. In 1988 ETSI was created and the power of standard regulation was transferred to from CEPT to the newly created organisation. In 1992 postal and telecommunications operators created their own organisations[265] and CEPT became a body of policy makers and regulators.[266]CEPT conferences became the forum were administrations, regulators and the industrial sector have been meeting to discuss issues of policy making in telecommunications.[267]After this substantial transformation, CEPT became the "umbrella" for the other three institutional bodies, ERC, ECTRA and CEPR.[268]CEPT's tasks have involved more substantial duties than in the past decades, such as for instance, discussion of sovereign and regulatory issues among representatives, assistance among the Member States, facilitation of relations among EU regulatory authorities, reduction of hierarchical and bureaucratic barriers within the CEPT Member States.[269]In 2001 CEPT was again reorganised in order to acquire a more active role as a forum of strategic planning and decision making. In addition, the two subcommittees, ERC and ECTRA were merged to one, the Electronic Communciations Committee (ECC).[270]

---

[265] CERP and ETNO.

[266] Andreas Evagora, CEPT Radio Conferences, Communications Week International, July 20, 1992, p.4

[267] See "CEPT Vade Mecum", September 1994 and The institutional framework for the regulation of telecommunications and the application of EC Competition Rules, Final Report for the European Commission, 1996. The relationship of the Commission and CEPT/ERC has been set on a political and a contractual basis and has acquired a character of co-operation, in which the EC plays an advisory role. The European Commission, with the support of the Council, co-operated closely with CEPT/ERC, especially on issues of S PCS frequencies and licensing. The Council showed preference to CEPT because of the following reasons. First, national interests could be better protected, which meant that the issue of competence distribution and national sovereignty among European Commission and Member States could be relaxed. Second, it would be a better solution in terms of technical capability and third, CEPT had a wider coverage of countries than the EC. The Council invited the Commission and the Member States to rely in the first instance upon the expertise available in the ERC/ETO and to co-operate with the ERC and the ECTRA. The Commission signed a MoU with ERC, a framework agreement with ERO, and a MoU with ECTRA.

[268] ECTRA examined and developped common telecommunications regulatory policies in a European context, taking account of European and international legislation and regulations. Furthermore it prepared European common positions for use when dealing with international bodies and exchanges views on issues of common concern to regulatory authorities in the CEPT Member Countries. Finally it establishes and elaborates regular contracts with representatives of relevant bodies within or outside CEPT. CEPR deals with issues related to postal services.

[269] For more information of CEPT, see www.cept.org.

[270] On the duties and responsibilites of ECC, see http://www.ero.dk/ecc. This study will concentrate on mainly on ERC, which has played a very important role for the policy harmonisation and regulatory introduction of S PCS.

b. CEPT/ERC

ERC has been one of the regulatory institutions that have played a key role in shaping a harmonised regulatory policy for frequencies before being merged with ECTRA to ECC; it has been the forum, where spectrum management in Europe has been taking place. It constituted a representative example of conflicts of interests between public and private actors at a European level, however, planning of harmonised pan-European spectrum policy has been efficiently taking place.

ERC has become the key institution for the harmonisation of spectrum policies: through a series of important documents; procedures were established that lead to the elaboration of rules for the creation of a policy for the introduction of S PCS. The analysis of these documents illustrated the relations between the regulatory bodies and the interest groups and as a consequence justified the nature of the resulting rules and procedures.[271]

ERC was a body composed of 4 permanent working groups, which created project teams to deal with defined tasks in a limited time.[272]In the framework of these projects various interest groups met: it hosted European multi-billion industrialists and important safety-critical public services, as well as public administrations, on whose proposals the ERC and CEPT decisions relied, under the constant participation of other interested groups. The model on which ERC decisions should be based, had to take into consideration certain important aspects. First, decisions should address major issues of pan-European significance. Second, they should be definitive in nature and not offer a menu of options. Third, they should have wide-spread support and be implemented by most, if not all, CEPT Member Countries, and particularly by EU Member States. Fourth, issues, priorities and time scales should be determined by close liaison between CEPT, EU and ETSI and fifth, decisions should give a clear signal to industry on CEPT/ERC policy.

ERC, besides tracing the common harmonised European policy in frequencies, had also been the mechanism for working on European Common Proposals at

[271] This framework has been formed from a series of documents released by the Commission, with the support of the Council, where the importance of ERC law production has been recognized. This has been the "point of reconciliation" between the supranational Commission and the Member States, which preferred an intergovernmental body to deal with issues such as frequency regulation. This approach, as elaborated by the Commission in 1993 and 1995, was re-confirmed by the Council in 1997, through the enactment of the S PCS decision. See chapter V for more details.

[272] These working groups were on Frequency Management, Spectrum Engineering, Radio Regulations and Preparatory Working Group.

the WRCs.[273]The reasons for using ERC as a means to promote international policy objectives were various. ERC has been more effective in producing proposals internationally in order to promote European initiatives, and at defending European interests, in particular in international fora, where Community issues had not been well represented.

The ERC decisions were developed in the light of promoting harmonised European policies. They focused particularly on the following aspects. First, facilitate the free circulation of terminals. Second, create a regime of general licenses. Third, create harmonised licensing conditions for gateway service providers, space segment and typed approved equipment. Fourth, provide GSM interworking capabilities and fifth, create a general framework in order to facilitate mutual access to markets around the world.[274]

c. Committees of expertise based on ERC/ECTRA regulation: The Milestone Review Committee (MRC)

The introduction of new telecommunications systems based on new technologies has led to the development of new bodies of expertise, the MRC being one of them.[275]The MRC has had no legal personality like the Licensing Committee, which was created through a Council directive. The MRC was established by ERC and ECTRA decisions, which were not legally binding [276]It has been an independent committee, composed of officials from the administrations that were signatories to the above mentioned ECTA decision, or that have committed themselves to implement the ERC decision.[277]Moreover the MRC has been composed of representatives of the ERO, ETO and the European Commission, who have an advisory role. The MRS has been closely working with the European Commission and the National Administrations.

---

[273] In particular a specific group, the CPG (Conference Preparatory Working Group) is responsible.

[274] See ERO, input into the preparatory meeting for the ITU "First World Telecommunications Policy Forum", 2 February, London, 1996.

[275] Scherer, Innovationsoffenheit der europaeischen Telekommunikations Regulierung-Konzepte und Instrumente; Das Beispiel der S PCS-Systeme, as above.

[276] These were the ERC/DEC/97(03) on the harmonization of frequency bands for the introduction of the S PCS and ECTRA/DEC/97(02) on the harmonisation of authorization's conditions for S PCS. ERC and ECTRA are both subcommittees of CEPT and have an intergovernmental nature. Their decisions become legally binding only if the Member States commit themselves to them.

[277] The first meeting of the MRC was held in October 1997. At that time, 16 countries from the 43 CEPT Member States committed themselves to implement the two decisions that created the MRC. These are: Norway, Germany, The UK, Sweden, Croatia, The Netherlands, Switzerland, Portugal, Liechtenstein, Finland, Spain, France, Lithuania, Italy, Belgium, Ireland.

The purpose of the Milestone Review Committee has been to facilitate the licensing process and to distinguish between real and paper S PCS systems.[278] In particular, according to the ECTRA decision, the MRC has had the following tasks. First, to monitor the compliance with the milestones by applicants in a transparent, non discriminatory manner and in accordance with the procedures established. Second, to seek information from the applicants on the compliance with the milestones. Third, to make recommendations to the administrations, ERC, and ECTRA, upon the examination of applications on the compliance with the milestones by S PCS systems and to handle any difficulties arising, in an appropriate manner in order to assist administrations in offering authorisations. Fourth, to make any modifications, and to finalise the new S PCS systems proposed by CEPT administrations that were intended to be brought into operation before January 1st, 2001.

In particular, as regards spectrum, the tasks of the MRC have been the following (according to point 1.5 and 1.6 of the ECTRA decision):

- to establish a procedure within the CEPT, including national administrations so as to monitor the spectrum used by the S PCS, for those systems that have been identified by the ITU as real S PCS systems and make recommendations to the national administrations, ERC and ECTRA if necessary and
- to produce reports for the national administrators, ERC and ECTRA on whether the scarcity of spectrum will be an obstacle to the number of the S PCS systems that can be provided within the CEPT and propose measures to overcome the difficulties.[279]

***II. Conclusions***

The regulators have been institutions at international, regional or national level whose objective has been to provide the legal framework, which could correspond to the demands of the interest groups. By taking into consideration the new technology, the needs of the actors and the specificities of new entries, they have tried to provide a regulatory framework that would cover their needs and shape the function of the market. We have seen that the more the regulatory level has moved from national to transnational, the more the number of partici-

---

[278] After the ITU allocated the new frequency bands for S PCS systems, the quantity of the new candidate systems was a problem, due to the lack of spectrum. In fact, more than fifty S PCS appeared as future providers of mobile personal communications. The ITU wanted to distinguish S PCS that wanted to provide services and operate, from systems that were only declared by name- the so called "paper"satellite systems-, wanted to get a frequency, but had no intention to become operational.

[279] The MRC has produced reports on licensing, as well as various recommendations as to whether or not there was compliance of Iridium, ICO and Globalstar with the Milestone Review Criteria. See for more, http://www.cept.org.

pants has been increasing and the more difficult it has become to deal with the increased number of actors and their needs, making conflicts of interests unavoidable. In particular, the most complicated web of interactions has been developed in the ITU because of the big number and the different character and interests of actors (national representations, international organisations, industrialists) as well as its global character as an institution.

It is interesting to see how rules have been made and decisions have been taken under the influence of industrialists, based on various lobbying techniques and negotiations procedures, as national delegations and industries have been urging the regulators to support their aim to getting more frequencies. The regulation that has emerged has been the result of many policy-making efforts as well as many negotiations and meetings with the economic actors. The relations developed between them have been of co-operative nature, which might have been co-decisive, advisory or observatory. Advisory has been, for instance, the role of the European Commission in the ERC meetings and observatory the role of the European Commission in the ITU conferences. On the basis of this schema of interaction as sketched out in this chapter, the next chapter will present and analyse the policy and legal framework that has been produced.

# Chapter V
# Building S PCS regulation

## *I. Introduction*

Law-making is a complicated process, for many a continuous one, for others partly evolutionary, partly static, which defines limits of human behaviour in societal structures, the market being one of them.[280]This process is based on law's dual functionality, as presented by the legal theory:[281]either law emerges as an expression of power, being one of its instruments, or law is a precondition or limit of political power. The first one had been the case in the telecom market until the beginning of the 1990s, where the provision of regulation constituted a dimension of political power for governments and administrative power for the National Telecommunications Organisations. However, during the 1990s, law became the regulatory instrument for limiting administrative and political power and shaping markets that integrate globalised economies.

Technologies, the behaviour of market actors and market function have always been a challenge to law. The elaboration of the legal framework has formed a process, the "reglementary process". This process has been an attempt to organise and control behaviour through the use of explicit rules.[282]In the S PCS case, for example, the idea was to adapt the already existing rules to the new technological, economic and political transformations by producing, where necessary, new explanatory or complementary rules. As a consequence, the legal framework had to be redesigned at a certain point so that it could adapt to new needs. This redesign of the legal rules has been based on the interaction among individuals and institutions, which has taken place on a number of different layers and in different levels of interrelation, such as dependence/autonomy, hierarchy/market based and replication/diversification .

The legal rules that have emerged are a dependent variable: the "legal output" of "policy outcomes"[283]constituting embodiments of policy decisions. Their success has been measured in terms of the extent to which they ensured that the substance of policy is achieved. Law describes how the policy outcome, which, in this case is the regulation, is a pattern or structure that emerges in a socio-po-

---

[280] See Jean Carbonier, Droit Flexible, 7e ed. Livrairie Génèrale du Droit and du Jurisprudence, Paris 1992, John Commons, A sociological view of sovereignty, New York, Kelley, 1967, p. 3.

[281] La Torre, Diritto, potere, dominion: Argomenti per una teoria non prescrittivistica del diritto, EUI 1992, Working Paper 92/30.

[282] Sally Falk Moore, Law as a process: an anthropological approach, London: Routledge and K. Plaul, 1978.

[283] Reasons and Rich, The Sociology of Law, Toronto: Butterworths, 1978.

litico-economic context as result of the interaction among actors, who all try to intervene, and where the recognition of dependencies is a dominant feature.[284]

Elements such as «flexibility» and «specialisation» have played a very important role in shaping the form of the legal rules. Flexibility has been a hallmark of a sophisticated regulatory regime, leading to the promotion of innovation and international competitiveness.[285]This has been possible in two ways: either through changes in the status of any detailed rules that might exist or through the use of general principles and vague rules like the obligation to provide a "safety net". In fact, flexibility as a regulatory tool has performed two functions: on the one hand it has "protected" the regulators from failing to regulate a new market and on the other it protected the market from being regulated too much or from being regulated in a way that would lead to market failure.

This process of generalising law and making it more flexible has been characterised as "proceduralisation" of law. This is a function, which balances the way in which regulatory power is exercised and how decision-making processes are organised. Furthermore, it helps administrative institutions to become more flexible and organise their capacities better. As Colin Scott describes:[286]«the process of proceduralisation of telecommunications law has three facts [...]. First there is a shift from highly technical and detailed rules towards reliance on more general norms. Second, there is greater emphasis on the procedures under which decisions are taken. Third, there is a shift towards the use of more general regulatory institutions and the practice of telecommunications regulation».

Specialisation has become another dimension of knowledge.[287]In the era of advanced technology, specialisation in terms of information, scientific expertise and know-how has become a necessity for the elaboration of new regulation. Proof of this new phenomenon has been the creation of various scientific bodies that assist regulators at international, European and national level to enact legis-

[284] This is one of the approaches-models of governance, which was developed by Kooiman as the socio-cybernetic system, in Kooiman, 1993a, Socio-political governance: introduction, in Kooiman, Modern Governance.

[285] Julia Black, Rules and Regulations, 1997.

[286] See Colin Scott, The proceduralization of telecommunications law, Telecommunications Policy, Vol. 22, No.3, 1998, 243-254.

[287] On specialisation, knowledge and information society, see Fritz Machlup, Knowledge, Its creation, distribution and economic significance, 1980. Technology has been defined as "the knowledge to design goods and services to obtain certain objectives". See J.W. Hans Markvoort, Government Regulation of Information-based product markets-The case of computer software, Diss. 1995, p.24.

lation,[288] as technical and scientific expertise has had to be provided in all stages of law generation and law application procedures, in order to facilitate co-operation between the community and national administrations.[289]

**S PCS specific regulation**
The emergence of the S PCS technology in the late 1980s and its use by the strategic economic actors who were, or got, involved in the satellite sector has been a new challenge to regulators. The introduction of the new systems with global coverage and the economic activities that went far beyond national borders, and the regulatory needs, needed coordinated and harmonised national regulatory approach. As a consequence, the necessity for the elaboration of a legal framework to deal with issues related to administrative, economic and legislative questions was made clear, not however in order to add heavy and more sector specific legal rules but to coordinate divergent regional regimes. These included the assignment of new frequencies and the elaboration of new licensing procedures, to give but one example.

Satellites were one of the first sectors where a liberalisation' basis with regard to the service provision was established, within the general trend towards liberalisation of the telecommunications sector. Satellite-specific regulatory policy has been taking place since the release of the Satellite Green Paper in 1990, whereas specific PCS regulation only started in 1997.

As a consequence, the regulation relevant to S PCS comprised different categories, such as regulation on general satellite communications, sector-specific regulation on frequencies and licensing, and S PCS specific regulation related to various economic and political issues, for instance, telecoms liberalisation marked by the loss of privileges by the NTOs, issues of frequencies and licensing related to state sovereignty.

The chapter will focus on the structural elements of the law that have been governing the market function as well as on the procedural requirements that have been giving rising to. Issues of law-generation and law-application examined in this study are related to soft law, market formation (or re-formation) and legal

---

[288] For example, at a European level, among the committees that have been assisting the Commission in regulating issues of frequency assignment: Open Network Provision Committee, European Telecommunications Committee, European Space Agency, European Conference for Post and Telecommunications, European Telecommunications Committee, European Standardisation's Committee, European Telecommunications Standardisations Institute, Milestone Review Committee.
[289] Dehousse Renaud, Joerges Christian, Majone Giandomenico, Snyder Francis, Europe After Maastricht, New Regulatory Strategies, EUI, Working Paper No. 92/31, p. 19.

and administrative co-operation.[290] The chapter will present and analyse first, the policy and regulation on satellite communications, which is related to S PCS, such as the Spectrum Green Paper and the Satellite Directive and will then analyse the policy and regulation on S PCS, such as the S PCS Decision.

## *II. International Level- ITU and S PCS regulation*

### 1. S PCS Related procedures

Before launching a satellite system, the administration which has awarded the space segment license to the project, has to go through a series of administrative procedures, which may last up to six years.[291] A satellite S PCS is necessary to apply the provisions of the ITU Radio Regulation with regard to advanced publication, co-ordination and notification. Advanced publication is the first procedure, which has to be applied both for GSOs as well as NGSOs and should start two to six years before the satellite network is brought into service. The objective of the advanced publication is to inform other administrations as early as possible about plans for new satellite networks, as this enables other administrators to identify possible interference problems with other existing or planned satellite networks and/or terrestrial networks.[292]

The second phase is the one of co-ordination, during which the proposing administration can operate the system, but only on a non-interference basis. In this phase, no earlier than six months from when the procedure of co-ordination is initiated, the administration that is proposing the system may file a formula called AP3, which includes specific information data so that other administra-

---

[290] An example of legal cooperation is the cooperation between the European Commission and the NRAs , when a systems applies for an operational license for the first time in one of the Member States of the European Union. The cooperation between the NRAs and the MRC, as to whether or not the applicant operator follows, according to the relevant decisions, the procedures for the granting of a license, as well as co-operation among the relevant administrations in the case of an OSS (One-Stop-Shopping) constitute examples of administrative co-operation.

[291] For instance, the UK for ICO, the US for Orbcomm, Globalstar, Iridium and Teledesic. The whole process of co-ordination is a process of internal notice and negotiation that facilitates the harmonious operation of in-orbit satellite networks, allowing no unacceptable interference. Also ERO, Mobile Satellite Services, Applications, Final Report, 1995, p. 23.

[292] The administration should provide information about the space station and its intended service area on earth and the basic radio parameters of the associated earth stations and or those of the typical earth stations. ERO, MSS, 1995, p.40. The Advanced Publication is the submission of "Appendix 4" information by the proposing administration. This submission describes the satellite systems in general, including frequencies, power levels and orbital features. James M. Talens, Satellite Co-ordination, An introduction, Via Satellite, p.92, April 1998.

tions can evaluate the potential for unacceptable interference.[293] After that, the ITU submits a request for notification (the phase of notification) in order to list the systems to the ITU's Master Register. The co-ordination procedure requires the administrations to supply data about the MSS space stations as well as data about the associated feeder link station and the user terminals. The above-mentioned procedures have to be carried out by the administration of the country where the satellite organisation is based. In the case of an international organisation, this administration acts on behalf of the group of the named administrations.[294]

Moreover, with regard to S PCS, WARC-92 adopted in the Resolution 46 a recognition of the sovereign right of countries to decide if they want to participate in non-GEO systems and to determine the conditions of access to such systems from their territory. This resolution was adopted in order calm national fears that the global communications offered by S PCS and the nature of these systems would threaten their national sovereignty. According to the resolution, international registration of a global non GEO MSS system does not mean that the operator of this system is free to provide services everywhere. Thus, implementation of the new systems in the allocated bands should be done under two conditions. The first has been related to the right of each ITU Member State to decide how or whether to participate in these systems and to determine the terms and conditions for the access of the interested parties to such systems from the national territory. This meant that Member States were free to exercise their national sovereignty, determining the rules of the game for all S PCS market entrants and thus being able to protect their home markets. Particular importance seemed to be given by the resolution to the uplink transmission, the signals originating country, if one took into consideration the phrase "from its territory".

The second had to do with the obligation of entities and organisations to provide international or national telecommunications services for NGOSS, in order to operate at the point of delivery under the legal, financial and regulatory requirements of the Member State in whose territory these services are offered. It required that all new providers of S PCS services and operators should deliver services to this country, respecting the legal, financial and regulatory require-

[293] The AP3 formula is published and given to all other administrations, which have six months to respond.

[294] For example in the case of Inmarsat, it has been the administration of the UK acting on behalf of the signatories of Inmarsat. For the LEO participants, additional restrictions appeared in Section 2.13 of Resolution 46 (WARC-95) "administrations with which co-ordination is sought, as well as the administration seeking co-ordination, shall make all possible efforts to overcome the [technical or operational] difficulties in a manner acceptable to the parties concerned". Talens, as above.

ments of the country. This statement could be interpreted in two ways.[295] If "deliver" was interpreted as the government's authorisation for the provision of services in the home market, the conclusion would be the same as it was with the first condition. Another interpretation would be that an operator of an S PCS should operate in all territories served in order to remain faithfull to the rules laid down by the national government who authorised the operation of the satellite system, namely the rules that would apply equally to the provision of services within the national territory as well as the services provided outside the national territory of the government authorising the provision.

**2. Regulation for Satellite Broadband Systems (Giga LEOs)**

Prior to WARC-95, no one had ever though that Fixed Satellite Services (FSS) could also be offered in orbits other than the geostationary orbit.[296] It was WRC-95 which, for the first time, identified the new 18.8-19.3 Ghz (space-to-earth) and 28.6-29.1 GHz (earth-to-space) bands for use by non-GSO systems in fixed satellite services. It was a new approach and new option for the new broadband multimedia non geostationary satellite services. An additional 2x100 allocation in the 18.8-18.9 GHz and 28.6-28.7 GHz bands was decided during the WRC-97. Furthermore, power limits in most parts of the Ku and the Ka band were adopted in order to protect GSO systems and also NGO FSS from interference by other non-geostationary FSS systems.[297]

**3. Memorandum of Understanding (MoU) on Global Mobile Personal Communication Systems (GMPCS)**

This MoU was concluded in order to secure future personal communications. Relevant discussions were initiated during the World Telecommunications Policy Forum of 1996[298] among both ITU Member States and firms, as participants with the objective of facilitating arrangements for type-approval, marking, licensing, provision of traffic data and customers recommendation. The arrangements

---

[295] KPMG, Report on the introduction of S PCS in the European Community, 1994.

[296] See José Albuquerque, Thomas Leo Hayden, WRC-97 Completed Regulatory Framework Enabling Non-GSO FSS Systems in 2x500 Mhz of Centimetric Waves (Ka-Band), Global Communications, Asia 1998, Hanson Cooke.

[297] From the 2x500 MHz, the 500 were allocated to Teledesic, since Teledesic was the only project which met the requirements. Furthermore, the power limits were not introduced for the bands assigned to Teledesic, which means that only one systems may operate and that sharing is impossible.

[298] The World Telecommunications Policy Forum (WTFP) was established by the 1994 Kyoto Plenipotentiary Conference. It has been a forum where discussions take place on important issues in order to form policy lines. During the WTPF of 1996, the discussions focused on mobile personal communications and especially Big LEOs. The meeting has been attended by 833 delegates representing 128 Member States and 70 sector Members. For more information on the WTPF, see www.itu.int/itudoc/osg/spu/fora/45500_ww6.doc.

were finalised on October 7th, 1997 by the national administrations of 74 countries, ITU Member States, GMPCS operators, providers, and terminal manufacturers.

The idea was to establish a simple authorisation regime instead of the sanctioning of individual licenses. The importance of this agreement was laying in the fact that the signatories recognised that GMPCS would be specific systems that would operate under specific frequency assignment procedures, which had to be in accordance with ITU co-ordination procedures. Furthermore, mutual recognition and creation of a single procedure for type-approval regimes and marking procedures had to take place. Moreover, it was necessary to create a single licensing regime and it was stated that the GMPCS terminals should be exempt from individual licence. Finally, GPMCS operators and providers would be submitted to the national laws of each country, as well as the custom duties.[299] The agreement set the basis for the world-wide facilitation of circulation of the terminals. However, as the agreement was not legally binding, it could not be forced on the Member States, which meant that in the end the operator would have to go through the procedures of each national state. As a consequence, the MoU remained a text of general guidelines theoretically applicable, as firms still had to go through the bureaucratical national type-approval procedures existing in the Member States.

In particular, the MoU consisted of six articles. The first referred to the type-approval of terminals. According to this article, the signatories should mutually recognise essential requirements for the type-approval of terminals. Article 2 defined that licences should be granted on a basis of general licences. Article 3 referred to the marking of terminals and Article 4 to customs arrangements: recommendations should be developed in order to exempt GMPCS terminals from customs restrictions when brought into a country on a temporary or transitory basis. Article 5 dealt with access to traffic data and defined that a GMPCS operator should provide appropriate data concerning traffic originating in, or routed to, its national territory on a confidential basis to any authorised national authority which so requests. Moreover, it pointed out that a GMPCS operator should assist it with any measures intended to identify unauthorised traffic flows therein. Article 6 underlined the need for periodical review of the results of the MoU. [300]

---

[299] GMPCS-MoU Arrangements, ITU News 10/97.

[300] ITU adopted all proposals which the European Commission made, in particular as regards type-approval procedures: the Commission proposed that global mobile personal communication systems should not cause any harm to networks, users or equipment, since the objective was to be legally placed in the market and legally used. Moreover, the MoU adopted the pro-

### *III. Evolution of telecommunications (satellite) policy and regulation in Europe*

The creation of a regulatory framework for S PCS has been the result of a long, gradual process, characterised by steadiness and complexity and at times slow.[301] The objective of this policy, which started to be developed dynamically after the mid 1980s, was to design and set the regulatory framework for the telecomms and the information society. This would be realised by taking gradual steps, looking far beyond the technologies in existence at that time and expecting a considerable evolution of the economic importance of the sector.[302] As a consequence, the first efforts concentrated in forming a policy to open the tele-

---

posals of the Commission on issues of user safety, electromagnetic compatibility, and the effective use of radio spectrum.

[301] See Maria Michalis, European Union Broadcasting and Telecommunications. Towards a convergent regulatory regime?, European journal of Communications, v.14, n.2, June 1999. Also for an analytical presentation of the EC's telecommunications policy, see Scherer and Bartsch, Introduction, in Scherer (ed.), Telecommunications laws in Europe, 4th Edition, 1998.

[302] Grewlich, Europa im globalen Technologiewettlauf, Gütersloh: Bertelsmann, 1992, p.82 ff, Syed Kamall, Telecommunications Policy, Cartermill Publishing, 1996, p. 27 ff. Scherer, European Telecommunications Law: The Framework of the Treaty, in European Law Review, 1987, p.354 ff. For the history of the evolution: since the 1930s there have been various attempts to create a common zone in the postal and in the telecommunications area. However they have all been rejected, as issues of sovereignty were at stake. However, in the late 1970s telecommunications became an integral part of the rapidly growing industry, as they had started becoming independent from foreign investment and had showed that telecoms were much more than a public service sector. As a consequence, the European telecommunications policy evolved slowly during the early phase of the development of the EC. In 1980 the Commission and the Parliament identified market fragmentation as the main obstacle and issued a recommendation to the Council which called for harmonisation of the telecommunications sector, creation of a common market for terminal equipment and the opening of telecoms public procurement to Competition [COM(80)422]. After 1983 the Community started to take drastic steps with the creation of the Task Force Group on Information Technology and Telecommunications. In 1984, a policy framework was created and the Senior Officials Group on Telecommunications (SOT-T) was created, which in 1984 submitted a proposed programme of action: [COM (84)277: Progress Report on the thinking and the work done in the filed of telecommunications and initial proposals for an action programme]. Moreover, in the same year the first co-operation agreement was signed between the Commission and CEPT. See Herbert Ungerer, Promoting European Communications, The role of the European Community, in David Shorrock (ed.) European Communications, Technologies and Regulations of the Single Market, Blenheim Online 1989, p. 165, and de Guiroye, A., 1993, Community Policy on telecommunications services: The main phases from 1984 to Maastricht, XIII Magazine, No. 11 Office for Official Publications of the European Communicates, Luxembourg, pp.1, 3Warnery F., 1993, Telecommunications Services: Towards Liberalisation , XIII Magazine, No.11, Office for Official Publications of the European Communities, Luxembourg, pp. 6-10, Natalicchi, p. 40ff.

com market and create the ideal conditions for competitiveness, since the creation of an internal market was one of the main aims of the Community.[303]

The strategy that the Community followed has been based on a series of Community and non-Community legally binding and non-binding documents, such as directives and amendment directives, Communications from the Commission, time limited regulations providing exceptions in favour of Member States, CEPT decisions and recommendations.[304] These documents followed policies developed by policy papers with political character, namely the Green Papers, which set the basis for the elaboration of a regulatory framework. These Green papers were released at regular intervals, covering all sectors of telecommunications.[305] The first Green Paper of 1987[306] was followed by the Satellite Green Paper,[307] which, was in turn, followed by the Green papers on Mobile and Personal communications,[308] the liberalisation of the telecommunications infrastructure and cable networks[309] and also numbering,[310] convergence,[311] and spectrum.[312]

---

303 The internal market programme was initialised in the 1980s and was a programme that would contribute to the development of the competition of the European economy through criteria of efficiency elaborated through strategies of deregulation. It was seen as a special strategy of economic integration, aiming at the establishment of an open market, which would guarantee rights of access, a key feature in the development of the S PCS market, as the objective of the Community has been the creation of a pan-European, barrier-free market, including sectors such as energy and telecommunications. However, in shaping the legal framework not only in the telecommunications sector, but also in all sectors related to advanced technology, there has always been a tension in forming a technology-based policy, as on the one hand there were those who supported the free market ideology while, on the other, there were those who criticised it. The supporters of the free market wanted the diminution of the internal barriers and the inherent fragmentation of the internal market, whereas those who were cautious had constantly warned of the danger, stemming from the ever increasing dependency on American and Japanese technology and their vision focused on a more positive and interventionist approach.

304 See König, Anmerkungen: Telekommunikationsrecht im Europa, EuZW, 21/1998, p. 655 ff. König, EuZW, Heft 21/1998, p. 656.

305 See also Scherer, Telecommunications Laws in Europe, 4th. Ed., 1998. On the factors that have contributed to the expansion of the European Commission competencies, see Natalicchi, p. 20.

306 Green Paper on the development of the common market for telecommunications services and equipment, COM(87)290, 30.06.87.

307 Green Paper on a common approach in the field of satellite communications in the European Community, COM(90)490, 20.11. 90.

308 Green Paper on a common approach to mobile and personal communications in the European Union, COM (94)145 final, 27.04.94.

309 This Green Paper was released in two parts: Green paper on the liberalisation of telecommunications infrastructure and cable TV networks, Part I, Principles and timetable, COM(94) 440 and Green Paper on the liberalisation of telecommunications infrastructure and cable TV

These important telecommunications milestones were materialised through legal and non-legal binding documents the most important of which were the directives. The policies that the Community had thus developed in the telecoms sector had been "published" as Green Papers and then made concrete through legal documents: a clear case of law following the policy.[313]

Each of the following papers set the policy basis for the elaboration of further legislation.

- The Green Paper of 1987 was the first general Green Paper setting the framework for telecommunications policy, covering all sectors;
- The Satellite Green Paper that followed in 1990 set the basis for restructuring in the satellite sector, proposing the extension of the principles of the EU telecommunications policy, as expressed in the Green Paper of 1987, to the satellite area;
- The Mobile Green Paper was an attempt to liberalise the provision of mobile services and the freedom of operators and focused on a universal mobile telecommunications system. The Infrastructure and Cable Network Green Paper has been emphasised more in the provision of services through cable infrastructure;[314]
- The Numbering Green Paper accomplished the creation of a common basis for numbering policy, focusing on issues such as number portability and equal access;[315]

---

networks- Part II: A common approach to the provision of infrastructure in the European Union, COM(94) 682.

[310] Towards a European Numbering Environment: Green Paper on a numbering policy for telecommunications services in Europe, COM (96) 590.

[311] Green Paper on the Convergence of the Telecommunications, Media and Information Technology Sectors, COM(97)623.

[312] Green Paper on Radio spectrum policy in the context of the European Community policies such as telecommunications, broadcasting, transport, research and development COM(98)596.

[313] See Keith Spicer, Responses to globalization of telecommunications, Telecommunications policy, December 1992, p. 729 ff. Also, Ungerer, Comments on Telecommunications Regulatory Reform in the European Community, in Majone G., supra, who underlines that the telecommunications policy has viewed telecomms as a major issue in the general transformation of Europe's economy into a service-based economy.

[314] Part One of the Infrastructure Competition Green Paper adopted the principle that providers of the telecommunications services in the EU should have a free choice as regards the infrastructure it delivers these services. The second part of the Green Paper sought to initiate a debate on important issues such as how could universal service be developed, and how telecommunications networks could be licensed. See Syed Kamall, Telecommunications in EU, p. 36.

[315] See Green Paper on Numbering, COM(96)590.

- The Convergence Green Paper pointed out the need for the creation of a common policy and regulatory approach as technologies converge and initiated a debate for a common approach towards the multimedia market;[316]
- The Spectrum Green Paper tried to make the Member States realise the importance of spectrum for the development of telecommunications in Europe and the necessity of initiating a common framework for dialogue among the Member States;

The important Green Papers for the evolution of the S PCS have been considered to be both the 1987 Green Paper, as the green paper-basis for all telecommunications and the Satellite Green Paper. The Mobile Green Paper and the Spectrum Green Paper are also important papers.[317]The Mobile Green Paper has set the basis for a community policy in the field of mobile and personal communications, which comprise S PCS, whereas the Spectrum Green Paper initiated a debate on forming harmonised spectrum policy in the Member States.

The S PCS policy was part of the general satellite telecommunications policy, which started to take shape in mid-late 1980s. There has been a parallel evolution of the regulation on satellite communications and S PCS: The Satellite Green Paper, which formulated the Community-wide policy on satellite communications, was released in the Community in 1990. The first document, which introduced the concept of S PCS communications in the Community, was released in 1993 by the Commission, followed by a Council Resolution.[318]One year later, it was the Commission, which enacted the Satellite Directive that liberalised the provision of satellite services and, in 1995; the Commission released a further communication on S PCS.[319]Two years later the Council's "S PCS de-

[316] On the Green Paper on Convergence, see, Emilio Lera, Towards multiservice personal communications: perspectives for a sector structure evolution, Telecommunications Policy, vol. 20, n.7, 1996.
[317] The Mobile Green Paper has set the basis for a community policy in the field of mobile and personal communications, which comprise S PCS, whereas the Spectrum Green Paper initiates a debate on forming harmonised spectrum policy in the Member States. Spectrum is a very precious resource, as it constitutes the main precondition for the operation of all satellite systems.
[318] Communication by the Commission on the introduction of satellite personal communications, COM(93)171, Council Decision of 16.12.93 on the introduction of S PCS , OJ C No 339, 1.
[319] COM(95)529, Proposal for a parliament and council decision on an action at a Union level in the filed of satellite personal communications in the European Union.

cision", followed, demonstrating that the Community had taken S PCS into serious consideration.[320]

### *IV. Setting the legal framework for market entry*

#### 1. The barriers to entry

The market was expected to witness several types of barriers, i.e., factors that make entry to new comers unprofitable, as it was a new market and the active players were few.[321]This meant that issues like abuse of dominant position, price fixing, interconnection, universal service obligations were not at stake, but could have been if more systems became market operational. Important factors for entry have been the following: first, spectrum availability and pricing. Complicated and time consuming procedures as well as high spectrum prices made it possible for only some players to compete. Second, licensing for the operation of the network operation as well as for the provision of services.

#### A. Spectrum

The possession of the necessary resources, (spectrum technology and investment) has been the necessary pre-conditions of entry to the market. With regard to spectrum, economic actors have had indirect access to it. As national governments have had direct access to spectrum, through their signatories-the NTOs, operators have been obliged to ask for spectrum from the governments and the price they have been asked to pay has usually been very high.

There have been efforts to set the foundation for a co-ordinated spectrum policy among the European countries, based on a joint effort of the CEPT and the European Community. It is important to point out that in regard to spectrum - besides the directives on GSM, ERMES and DECT introduction released by the Council - the other documents related to spectrum have not been legal binding. This has demonstrated the difficulty in forming not only a legal, but also a policy and a framework for spectrum co-ordination, something which both CEPT and the European Community have pointed out.[322]The European Commission's Spectrum Green Paper and the ERC's numerous decisions have had the same

---

[320] Decision No 710/97/EC of the European Parliament and the Council of 24 March 1997 on a co-coordinated authorization approach in the filed of satellite personal- communications services in the Community, OJ L No 105/4, 23.04. 1997.

[321] Demsetz, Barriers to entry, The American Economic Review, March 1982, p. 47 ff., also, Geroski P., Gilbert R., Jaquemin Alexis, Barriers to entry and strategic competition, Chur: Harwood Academic, 1990, Yip, George, Barriers to entry: a corporate strategy perspective, Lexington Books, 1982, Wiiliamson Peter J., From barriers to entry, to barriers of survival, Fontainebleau: INSEAD, 1998, Working Paper.

[322] See ERC/DEC/(97)03 on the coordinated introduction of S PCS in the Community and ERC/DEC/(97)07 on the coordinated introduction of UMTS in the Community.

focus, neither of which indicate, however, how national states should assign their frequencies, (e.g, through auctions, lotteries or administrative procedures) nor pricing methods.

a. Marketing Spectrum

Spectrum is allocated at international level, allotted to national governments and then assigned by them to interested parties.[323]Once priced it is marketed. There have been various spectrum pricing mechanisms and discussions on which of them is most appropriate.[324]Spectrum pricing could be an obstacle to the market in two ways. First, if different systems are being used in the various countries and second, if prices are too high and the operator, who has bought a spectrum portion, has no other sources for the operation of the system. Pricing could also constitute an obstacle if interested parties would need too long to find information for the fees and conditions, which are not included in the frequency allocation tables.

Among the mechanisms, that of auctions seemed to be the method preferred by many administrations.[325]The issue of auctions had often been discussed. Those in favour of auctioning believe that first, auctions are quicker and more economical than administrative processes and are that they transparent, given that the rules of the auction obligate the government to be explicit about the criteria. Second, auctions are a flexible policy making tool. The countries using it have adopted a range of criteria for awarding licenses and the government can guard against monopolisation by imposing rules limiting the amount of spectrum -that a simple firm can hold- within a given geographical region. Third, auction does

[323] The term "allocation" refers to the entry of the new frequencies in the Table of Frequency Allocation in the Radio Regulations. According to the definition of the ITU it is "the entry in the table of frequencies of a given frequency for the purpose of its use by one or more terrestrial or space communication services." (ITU, RRs, Article 1, No 17(1982). After frequencies have been allocated to services (and not to countries), they are allotted to countries. It is then the task of each government to assign the allocated frequencies. The term "allotment" refers to the entry of a designated frequency channel in an agreed plan to be adopted by a conference and used by the Administration for a station to use a radio frequency or a channel.

[324] See NERA, Approaches to the allocation and assignment of the radio spectrum resource, Final Report for the EC, November 1997.

[325] See Milgrom Paul, Auctions and Bidding: A Primer, in Journal of Economic Perspectives, Summer 1989, 3, 3-22 and Milgrom Paul and Robert J. Weber, A theory of auction and competitive bidding, Econometrica, September 1982a, 50, 1089-122, Graham Daniel and Robert Marshall, Collusive Bidder Behavior at a single Object Second Price and English Journal of Political Economy. FCC has developed advanced methods for auctioning, having deviled the US geographically and the spectrum by wavelength, making thousands of licenses. See John McMillan, Why auction the Spectrum, Telecommunications Policy, 19, n.3, 1995, p.191-195, 151.

not lead to ownership, as the government is not auctioning spectrum but spectrum rights.[326]

However, the Satellite Action Plan Regulatory Working Group argued that auctioning spectrum could not be appropriate for various reasons, including: first, the uncertainty regarding the ultimate cost of the auctioned spectrum and the amount of time needed to finance the necessary arrangements could undermine investment in new satellite systems. Second, auctions could result in a reduction of competition, since the winner already had his assignment. Third, international satellite systems required access to spectrum on a global basis; ideally the same spectrum for all countries and an auction process could result in different assignments in different countries and could lead to inefficient spectrum use. Fourth, the total auction cost to a satellite would be the sum of the costs in all the countries where auctions are operated. This sum could exceed the cost of implementing the system, could make the system unattractive to investors and could result in an uneconomical service charge.[327]

**B. Licensing**

The various stages of licensing have comprised numerous actors, who often had to operate under different legal regimes. The system initially needed a space segment license, which would be granted only by one national administration, to which the system operator would apply.[328]The national administration has been the intermediate actor between the system operators and the ITU. It is this national administation that initiated the procedures of advanced publication, notification and co-ordination, on the basis of which all other administrations could be informed about the system's intension to become operational. In particular, with the advanced publication, the system had to submit all necessary technical and operational information, whereas with the co-ordination and notification the member states had to state if the operation of the system would cause harmful interference to their systems.

After the granting of the space segment license, the system had to obtain licenses for, first, the operation of the network in the countries, where infrastructure, for instance gateways, had to be installed. If space segment operators and network operators were the same entity, they would be the ones that had to do it. If not, then it was the network operators that had to proceed to the licensing. License, however, has not only been needed for the instalment of the net-

[326] McMillan, as supra. See on discussions on auctions,
[327] See SAPRWG, Report on Market Access, 1998, p. 63.
[328] Examples of such administrations have been: the US, which granted the license to Iridium and Globalstar, and the UK to ICO.

work infrastructure, but also for the provision of the services. This license has been the responsibility of the service providers.

The licensing of the network operators and the service providers has been subjected to the competence of the national government where the system wanted to operate. Existing procedures have been strict and related to factors sometimes disappointing for the market participants, for instance very high licensing fees. The lack of such procedures by the national administrations have been two common phenomena that caused unnecessary delays and loses of money for the operators.

a. High licensing fees for satellite earth stations and for service providers, high customs duties on equipment

For equipment to be taken into a country either on a temporary basis or for import, customs duties have been added to the cost of equipment for users or operators and these reduce the potential market access. Furthermore, fees have been rather high. However, the Licensing Directive suggested that fees imposed by administrations should seek only to cover the administrative cost incurred, that is, for management, control and the enforcement of the applicable individual licenses (Article 6). Fees should be proportionate to the work involved (Article 11, 1). In addition, Member States have been free to impose charges that reflected the need to ensure the optimal use of these resources (Article 11.2).

Costs have usually been divided into the following categories. First, initial administrative costs linked to the processing of individual applications. These costs arise only in the first year and depend on the importance of the network. Second, costs of co-ordination procedures at national and international levels. These costs occur only in the first year and depend on the number of stations. Third, permanent administrative costs linked to the management, control and enforcement of the authorisation scheme. These costs are yearly and should mainly be imposed on individual licensees. Fourth, costs related to the use of frequency spectrum.[329]

The fees for granting a license may be set to recover the administrative costs of issuing the licenses and be a good source for earning new revenues for governments. The level of license fees may also depend on the length of time that a visitor operator is in a country.[330]There are various assumptions on the sum of

---

[329] On fees see European Commission, Fourth report on the implementation of the Telecommunications Regulatory Package (COM(98)594 and Fifth report on the implementation of the telecommunications regulatory package, (COM(99)537).

[330] See David Wright and Shola Taylor, A policy and regulatory framework for the mobile satellite services, Telecommunications Policy, vol. 20, No. 8, 1996, pp.549-555.

the costs for the licenses fees, which depend mainly the level of governmental intervention and the willingness of the government to introduce competition or not as a basis.[331]

**2. Satellite Green Paper**

Setting a policy framework to regulate satellite communications has not been an easy issue. This policy started with the Satellite Green Paper and culminated in the S PCS decision framed by many documents of a non-legal binding nature. Legislative efforts have focused on attracting the attention of the industrialists, and making their participation in the market more intense. In addition, the efforts were designed to facilitate free circulation, complete the internal market, create a common space for telecoms and abolish the privileges of the NTOs.

The Green Paper aimed at establishing a general framework which would be able to embrace all possible future developments. This is why it tried to establish a new, more flexible regulatory system in order to foster the growth of satellite services in Europe, a sector characterised by the international dimension of services provided, by the lack of technological development and by the absence of a satisfactory regulatory framework in Europe.[332]The Green Paper wished to establish a common approach in the field of satellite communications in the European Community. The Commission established principles on the liberalisation of the sector, and the effort was confirmed by the Council in 1993.[333]

**A. The first steps: pointing out the problems and aiming for restructuring**

By the early 1990s the Commission had already identified satellite communications as one of the important sectors for future development and was one of the first sectors in which the Commission decided to initiate liberalisation, pointing out the weakness the of European satellite industry vis a vis other competitors,

---

[331] A) Exorbitant costs: these costs have the role of financial re-compensation. The sums are various: they may be high, low or exorbitant and depend on the requirements of the government, b) high costs: these costs "express the intent", which means that the company is assumed to have all resources and the costs can be high., c) reasonable costs: the functionality of these costs is to cover the administrative costs for spectrum allocation, d) minimal costs: it is only a minimal sum, mainly in competitive environments and the sum depends on the government , e) no cost: the existence of no administrative fees is possible in two cases. First, if the competent NRA think that the procedures of collecting fees will bring less revenue to the state and second, if the NRA does not want to get involved in issuing individual licenses and grants a class license to a group. Also, World Wide Satellite Communications, Services and Regulations, Report to clients, introductory volume, September 1991, p. 29.

[332] See Malanczuk Peter, Ten years of European Telecommunications, Telecommunications & Space Journal, 1994, p. 41.

[333] Council Resolution of 7 December 1993 on the introduction of satellite personal communication services in the Community, OJ C No 339, 1, 16.12.93.

in particular the USA. As a consequence, the Commission decided to capture the attention of all interested parties by releasing the Satellite Green Paper.[334] This paper constituted a very important text legally, as it presented the legal obstacles and the deeper economic and political problems that made satellite communications stagnant on the one hand and incarnated the visions of European regulators for liberalised communications on the other.

a. The economicopolitical context

In order to understand the value of this Green Paper, which constitutes one of the milestones of the European Community's effort for liberalisation, it is important to refer to the economical and political context in which it was created. The Commission had realised early that satellite communications was a key sector for the future of telecommunications, a sector were industry should be the leading power. As a consequence this paper was also aimed strengthen the European industry that was weak in comparison to its US-counterpart. Hughes and Lockheed Martin, the two US satellite manufacturers, had, for the previous decades been monopolising satellite manufacturing at a global level. Furthermore, the weakness of the European industry was particularly revealed with the creation of Intelsat.[335]

The paper had targeted particular categories such as industrialists, operators, providers, manufacturers, national administrations and NTOs, which had already started to lose their privileges as national dominators. Moreover, it constituted a call for dynamic participation and involvement in the satellite market, which, at that time, was strongly influenced by telecommunications market transformations in the US, the UK, Australia, Japan, and Canada.[336]

In its attempt to promote satellite-based communications, in particular, and wireless communications in general, the Commission had to take two serious factors into consideration. First, Member States of the EC, which were, at the same time, all members of the International Satellite Organisations (ISOs), operated

---

[334] Green Paper on a common approach in the field of satellite communications in the European Community, COM(90)490.

[335] INTELSAT was actually operating under the auspices of COMSAT, which was the first private American satellite organisation. However, the creation of Intelsat not only had negative influence for the European industry, it also turned out to have a positive effect, since it ended up warming up the industrial relations among the Member States.

[336] For the above-mentioned countries, the time of deregulation and privatisation of their telecom markets had already started. For example, for the US, deregulation started with the break up of the AT&T in 1983. In UK the phase of deregulation and privatisation of BT lasted eight years, from 1981 until 1989. In Japan liberalisation begun with the privatisation of the NTO. In Australia the market became competitive in very short time, because of the co-operation of the NTO with the efforts to liberalise.

satellites and had access to space segment capacity through the ISOs. The NTOs were in the privileged position of representing their governments in the International Satellite Organisations (ISOs) as national signatories[337] that were created in order to facilitate international telephony services by establishing global networks. The initiative was taken by national governments with the US as the leading power. The ISOs gave the NTOs the privilege of renting or selling space segment capacity to all interested parties. As a consequence NTOs became the exclusive providers of space segment, having attributed 80% of the rights to access in the 1980s.

Second, frequencies were co-ordinated by the ITU in which all Member States participated, although there was no harmonised European policy. All EC Member States, being signatories of the ITU, had exclusive control of the satellite frequencies. Frequency allocation and orbital access took place in the framework of the ITU conferences and negotiations. The Commission did not want to challenge the already established international regime, but tried to expand its competencies in many areas of the satellite communications within the limits of those international arrangements.[338]

b. The gradual steps of liberalisation as presented by the paper

The Satellite Green Paper opened the way towards satellite communications liberalisation by indicating the weaknesses in the existing system on which the legislative effort should concentrate. The paper made it clear that the Commission's approach to satellite communication was more liberal than to other kinds of telecommunications, such as voice telephony. The Commission did not want to challenge the above-described situation, that is, both the existing international agreements and the competencies of the Member States who were all committed to the international agreements, but proposed to extend EC regulatory principles to many areas of satellite communications.[339] Furthermore the paper criticised the mentality of the 1960s and the 1970s, when satellites were seen as a means of communication in addition to the terrestrial networks and not as an autonomous telecommunications means. With the above-mentioned approach, important qualities of satellites were neglected, such as their capacity to cover very big zones of the earth.

---

[337] The signatory has been the national entity which has been authorised by the government to sign on its behalf.

[338] Natalicchi Giorgio, Telecommunications policy and integration processes in the European Union, Diss, New York, 1996.

[339] Ibid., p. 138.

The major changes which the Green Paper proposed were the following:[340]

- Full liberalisation of the earth segment, where the receive-only terminals were subjected to type approval procedures when they were connected to public switched networks, while the transmit and receive terminals were subjected to licensing approval procedures, where it was justified. The main reason for this proposal was the forthcoming competition that the ground segment facility providers, (NTOs), would have to face by the VSAT providers. With the VSAT systems no connection to the public network was needed, since the VSAT had the function of a network.
- Full commercial freedom for space segment providers. With this general statement, reference was indirectly made to EUTELSAT and it was argued that a structural change was necessary so as to allow EUTELSAT to operate without being influenced by the NTOs. In this way, new providers, who are interested in obtaining space segment capacity, would be also able to contact and deal directly with the ISOs, without the intervention of the NTO. The distribution of space segment capacity would thus apply competitive, instead of arbitrary, bidding to all service providers.
- Harmonisation measures to facilitate the provision of European wide services.
- Free and unrestricted access to space segment capacity. In the system established by the ISOs and the national governments, the contracting parties, namely the national signatories, were favoured. Since the national signatories were the NTOs, they both set the rules of the game and played the game. By this approach, the Commission wanted to favour all providers who are not contracting parties. The Commission believed that the distinction between contracting parties and non-contracting parties who had to buy capacity from the signatories should be abolished and that the private sector, that is, the licensed providers for private satellite services, should have access and should also have the obligations of public service providers.[341]

---

[340] Also Scherer, Telecommunications Laws in Europe, 3rd Ed. 1995, Verhoef, P., 1993, Satellite Communications, XII Magazine, No. 11, Office for Official Publications of the European Communities, Luxembourg, pp.25-32, also, Verhoef, The role of satellite communications policy in the European Community, paper for the EC/EUTELSAT Satellite Seminar, Moscow 18-19 May 1993.

[341] I&T, European Commission, DG XIII, April 1995. This observation highlighted the two key issues of paramount importance as regards European Community-ISOs, that is, the issue "separate systems" and "direct access". Both INTELSAT and EUTELSAT contained Articles (XIV (d) and XVI (a) respectively), according to which any member country was prohibited from establishing public telecommunications via satellite if "significant economic harm" were caused to the relevant ISO. Furthermore, both ISOs require technical compatibility. The question has been when does significant economic harm start, in particular now, with the launch of S PCS, which as mobile communications, will be considered as "separate systems". On the issue of the separate systems, the Council released a Resolution in 1991, urging the EC Member States to develop effective, non discriminatory and accelerated procedures for the estab-

As a consequence of the Satellite Green Paper, a series of important measures were adopted among them: a directive introducing mutual recognition of type approval for satellite earth station equipment in 1993, a Commission directive on abolishing special and exclusive rights for the provision of satellite services and equipment in 1994 and a council resolution on access to frequency spectrum adopted in 1994.[342]

**3. The Satellite Directive[343]**

A radical change in the market came with the "Satellite Directive", which followed four years after the Satellite Green Paper. It was an attempt by the Commission to encourage European industrialists, such as Matra Marconi, Alenia, Aerospatiale, Telespazio, and EUTELSAT, to get more dynamically involved in the satellite communications business. Moreover, it was one of the Commission's "liberalisation's directives" based on Article 86.3 of the Treaty.[344]

First, the directive aimed at establishing a regime where private operators would be able to offer their services directly to the users, without the intervention of the national signatory, either as a negotiator between the new comer and the International Organisation or as a seller of the space or ground segment capacity. For instance, if Matra Marconi wanted access to space segment capacity from Inmarsat, it would have to purchase it from France Telecom, which was the French signatory to Intelsat. Due to the arrival of the new satellite providers in the telecommunications sector, it was important that they carried out conformity assessments directly with the international organisations, without the intervention of the national signatory. Direct access meant that the interested parties saved money because with the same capital required to be connected to the public network, in order to provide their services publicly, they could run their own

---

lishment or separate satellite systems. Moreover, the Proposal of the Commission to the Council for a regime of mutual recognition of satellite network services states that "the space segment used for the provision of satellite services can be either provided by one of the ISOs or by a "separate" satellite system, this being either space segment of a national satellite system or of a private satellite systems." The second issue has been that of direct access. As mentioned above, the ISOs are international organisations, built up on a commercial basis, which means that they market space segment access capacity. In the EC, the signatories are the NTOs, which, in this way, enjoy exclusive rights to provide space segment capacity. The priority of the Community has been to make NTOs open access to space segment capacity by establishing conditions for fair, non-discriminatory and transparent access.

342 See below in this chapter.

343 European Commission, Directive of 13 October 1994 amending Directive 88/301/EEC and Directive 90/388/EEC in particular with regard to satellite communications, 94/46/EC, OJ No L268/15, 19.10.94.

344 More on Article 86.3, see Scherer and Bartsch, Introduction, in Scherer (ed.), Telecommunication Law and Policy of the European Union, 4th Ed., 1998.

network without the intervention of the public operator, and have their interests better represented.

Second, the term "exclusive and special rights" of the national telecom operators was redefined, as the old definition, contained in the Directive 88/301/EEC[345] did not include all relevant cases. One of the main concerns of the directive was the abolition of such rights. These rights include the provision of the satellite network services and broadcasting provided by public or private providers. The term special and exclusive rights was redefined as rights granted by a Member State to a limited number of undertakings engaged in the importation, marketing, connection and maintenance of the telecommunications equipment. Furthermore, and as a result, the term "terminal equipment", was amended to include the satellite earth station equipment.[346]In simple terms, the directive in this way equates ground stations with common end-users terminal equipment, where no privileged rights can be exercised. In so-doing, the directive abolished the specificity of the ground segment market and abolished the natural monopoly of the NTOs.

The third issue that the directive touched upon was the issue of licenses. Satellite operators should have full freedom to provide their services in the whole Community, once they are licensed in one Member State. Additional regulatory restrictions, other than those dealing with frequencies and site co-ordination arrangements, required under the international commitments of the Member States, are deemed to be contrary to Article 49 (ex Article 59) of the Treaty. There were questions to be solved. First, how should the licence process take place, second, under what conditions, and third, in what way could an EC-wide system facilitate the distribution of licenses. Moreover, there was the spectrum scarcity issue. This included, first, the difficulty of establishing a uniform system in Europe, and second, the limitation of the number of licenses attributed due to the physical limitations of the spectrum.

At the time of the release of the directive, users had to obtain licences from each country of operation. In some cases, they had to have licences for each terminal, which could result in as many as 5000 separate licenses being used for one multinational's network. One step towards easier access to satellite capacity was the development of a four party licensing agreement, the "one-stop-shopping" procedure, signed between France and the UK initially and subsequently followed

[345] The "Terminal Equipment" Directive introduced the liberalisation of the market for terminal equipment: Commission Directive of 16 May 1988 on competition in the markets in telecommunications terminal equipment, 88/301/EEC, OJ L No 131/73, 1988.
[346] The satellite earth station equipment was not included in the definition of the term "terminal equipment" included in the Directive 88/301/EEC.

by the Netherlands and Germany. The result has been that since 1993 only one licence application needs to be submitted when the provider wishes to provide services in each of these countries. Nevertheless each country retains its own terms and conditions and licence fees are payable to each country. However, the distribution of licences, as regards terminals not connected to the public switched network, should protect privacy, whereas for the stations and other terminals connected, it should correspond to conformity assessments.

Three years after the Satellite Green Paper and one year before the Satellite Directive and the Mobile Green Paper,[347]the Commission released a communication, introducing S PCS in the Community.[348]The policy framework had already been set.

**4. Scepticism, and challenge to the market players: Satellite Action Plan(SAP)[349]**

Both regulators and industrialists were involved in the SAP. It has been a document which introduced a policy on developing and strengthening satellite communications, as it constituted a dynamic call for more participation in satellite communications from manufacturers, operators, and providers, whose involvement has been constantly increasing. The involvement of Loral in Orion, Boeing

---

[347] Green Paper on Mobile and Personal Communications: in Part 1994 (Toward the Personal Communications Environment: Green Paper on a common approach in the field of mobile and personal communications in the European union COM(94) 145 final, 27 April, 1994. The paper suggested: first, the elimination of all remaining exclusive and special rights in the sector, subject, where required, to the establishment of appropriate licensing conditions. Second, the removal of all restrictions on the provision of mobile services both by independent services providers and direct services provision by mobile network operators. Third, the facilitation of pan- European operation an service provision. Fourth, the full freedom of the mobile network operators to operate and develop their networks for the purpose of all activities provided for in their license. Based on the Mobile Green Paper, the Council and the Parliament issued a directive to abolish the special and exclusive rights in the mobile sector, amending the Services Directive. The directive provided for liberalisation of infrastructure, the right to direct interconnection, the possible right to request a license and operate digital mobile services using DCS 1800 technologies from 1 January 1998.

[348] Communication from the Commission to the European Parliament and Council on the introduction of S PCS in the European Community, COM(93)171.

[349] EU Action Plan: Satellite Communications in the Information Society, Communications from the Commission to the Council, the European Parliament, the Economic and Social Committee and the Committee of the Regions, COM (97)91 final, released on 5 March 1997. The new version of the Satellite Action Plan, which has been adopted as an official Commission documents, gave an overall presentation of the industrial and regulatory situation, emphasising on the recent developments on S PCS: see European Commission, The EU Action Plan, Satellite Communications in the Information Society, recent activities, present situation and outlook, SEC (1999) 250, 16.02.99.

in Ellipso and Skybridge, Alcatel in Skybridge and Matra Marconi in Teledesic have been some examples of this trend.[350]

The SAP tried to set priority actions to be undertaken at EU level, so that the appropriate interface with the industry would be secured. Furthermore, it signalled the ever-increasing importance of the sector. The paper aimed at five different sectors: the strengthening of the European industry, the completion of the internal market in satellite communications, the reinforcement of the European position at international level, and the reinforcement of the European Community R&D support.

First, action had to be taken so as to establish the proper environment to improve the industrial competitiveness of the European Union and to break the monopoly of the US in particular in the S PCS communications, as the majority of the S PCS announced were US lead, Iridium, Globalstar, Orbcomm, Teledesic, Astrolink, and Leo One to name but a few. Thus, the Commission called for co-operation between all the relevant actors, including industry, operators, and the Member States in order to strengthen the role of European industrialists and companies with world-wide power, such as Ericsson, Matra Marconi, and Alcatel, in global, advanced broadband, multimedia satellite systems, services and applications.

Second, the implementation of the measures for satellite telecommunications had to be rapid. The lack of implementation of legislative measures, which, in some cases, revealed insufficient regulatory framework as well as insufficient co-operation and co-ordination between national authorities, had as a consequence, the disintegration of national authorisation procedures and the existence of very long, bureaucratic administrative procedures. As a consequence, interested operators had to wait a long time and the granting of the license was not guaranteed.

Third, the Commission referred to the work of CEPT. CEPT's measures had become a basis for the implementation of Union regulatory policies. The decisions on spectrum harmonisation for S PCS and UMTS have been examples of this.[351] However the CEPT framework of intergovernmental co-operation and the voluntary adoption of its proposed measures had not proven to be sufficiently effective, as the decisions became legally binding only when the Member States committed themselves to them. Thus, the Commission requested CEPT to inten-

[350] See European Commission, The EU Action Plan: Satellite Communications in the Information Society, Recent activities, present situation and outlook, Staff Working Paper, SEC(1999)250, Brussels, 16. 2.99, p.5.
[351] Decisions can be found at http://www.cept.org.

sify its efforts to adopt the measures, particularly the frequency and licensing procedures, and to increase the efficiency of its regulatory decisions.[352]

Fourth, the European position could only be improved through certain actions, which include, first of all, the improvement of market access. This can be achieved through the implementation of the commitments of the Member States within the framework of the recently concluded WTO Agreement on Telecommunications[353] as well as by an in depth analysis of the trade policy regarding issues of international satellite communications.

Fifth, research and technological development would play a crucial role in the information society. Thus, the support of the development of generic technologies, the implementation of pilot networks and of innovative systems and manufacturing processes will be of great importance.[354] However, as the Commission pointed out, for the realisation of all the above-mentioned objectives co-operation between all interest groups (industry, networks operators, services providers, equipment manufacturers, content providers, users, consumers, national authorities, and European institutions) would be the most important condition.

With the SAP the European Commission has made an effort to come closer to the industry,[355] and has recognised the regulatory difficulties that industrialists have to face in terms of regulation and administrative procedures and has proposed a common effort so as to overcome these barriers.[356]

---

352 See Communication from the Commission COM(95)85.

353 Signed in Geneva on February 15th, 1997, See SAP, as above, p.18.

354 There have been various activities by ESA, NASA and the Community's 4th and 5th Framework program, such as ESPRIT and TEN-Telecom. See SAP, as above, p. 23 ff.

355 See David Wright, The European Commission and the Satellite industry combine forces to open markets, Telecommunications Policy, 1998 Vol. 22, no. 4/5, pp.267-272.

356 As a response to the SAP, a Satellite Regulatory Working Group has been set up to respond to the SAP and released a very interesting report called "Market Access: problems and solutions ", October 1999, where the following issues were pointed out: first, the need for effective and timely implementation of the CEPT decisions and recommendations, second, the need for timely and effective implementation of the EU directives and third, the need for the Community support to gain market access to third countries. Furthermore, the group has made has made 39 recommendations in order to help the Commission co-ordinate its regulatory efforts. Among them, for instance, the following are included: encouragement of the Commission for effective implementation of CEPT decision and Recommendations by all Member States, co-operation among all regulatory institutions at Community level, infringement actions against all Member States that have not implemented relevant directives, and abolition of the requirements for individual licensing for the future Broadband Multimedia Satellite Systems. See David Wright, The European Commission and the Satellite industry combine forces to open markets., Telecommunications policy, vol. 22, no. 4/5, p.267-272, 1998, also

**5. Licensing Directive**

The liberalisation process influenced the licensing procedures, by first, establishing fair, non-discriminatory and transparent selection criteria and second, by reinforcing co-ordination and co-operation among administrations in order to facilitate the operation of the systems among the various countries. Licensing has been the key issue for the S PCS systems, taking place under the general guidelines of the "Satellite Directive"[357] under which the regulatory authorities have been obliged to ensure market access to any authorised S PCS. One main feature of the licensing procedures has been that they can remain a regulatory tool through which public authorities can exercise control over their national markets. However, what the European Community has tried to do has been to simplify licensing rules and conditions, as complex procedures have been seen as an obstacle to the competition of a single market in telecommunications.[358]

Due to the wide divergencies that national regimes presented, the EU wanted to create a regime where the main feature would be the harmonisation and simplification of the procedures as well as the licensing conditions and the introduction of one-stop-shopping procedures (OSS).[359]The objective has been the creation of a simple regime, which would be attractive to the industrialists. The first steps were taken by the Commission and aimed at forming a policy for the mutual recognition of licences and other national authorisations in satellite communication services and networks and in telecommunications in general.[360] Mutual

---

Gerald E. Oberst, Jr., Benefits of open markets in satellite communications, Telecommunications Policy, vol. 22, no.9, pp.739-743, 1998.

[357] Accompanied by the Council resolution 710/97 and CEPT/ERC Decisions (03) and (04) of 1997.

[358] See Colin Scot, The proceduralization of EC law, Telecommunications Policy, vol.22, No.3, pp.243-254

[359] In Denmark, for instance, operators were free to enter the market without formality. In the Netherlands, the operator were required to register, while in Finland as well as in Sweden, they had to notify their intention to do so, except where the use of the frequency spectrum is requested. These are considered to be regimes with light licensing regimes. However, there are others, where individual licenses are the rule, and in some cases, the government minister is required to sign every license.

See Communication from the Commission to the European Parliament, the Council, the Economic and Social Committee and the Committee of the Regions, Fifth Report on the Implementation of the Telecommunications Regulatory Package, COM(1999)537, 10.11.99, p.12.

[360] COM(93) 652 proposal for a Council directive on policy for the mutual recognition of licenses and another national authorisation for the provision of satellite network service and /or satellite communications services. The objective of this proposal was the elaboration of balanced and efficient procedures for mutual recognition of licenses and authorisations for satellite services and networks. This would mean, that once an operator was licensed in one Member State they could operate in any Member State. The next proposal, COM(94)41, was a proposal on the mutual recognition within interim procedures based on one-stop-shopping li-

recognition meant that a level of protection of the harmonised essential requirements was obtained, and that this would be valid in every Member State.

The Commission wanted to change the regime where telecom operators (and satellite operators) that wanted to provide services in more than one Member States had to apply for a separate licence in each state of operation. For a satellite operator this meant that he would have to apply for an individual licence for each satellite earth station. These attempts went beyond ensuring Union-wide competition; they reflected the role of authorisation regimes in imposing rights and obligations and in monitoring markets.

According to the Commission, licensing should take place on the basis of the following principles.[361]First, the awarding authorities should be independent from the actors in the sector. Second, the procedures should be transparent and non-discriminatory, set against defined timetables, and the decisions should be open to appeal. Third, the fees related to the awarding of the license should be in proportion to the level of effort involved in administering the licensing process and not a levy of profits. In the case of radio frequency licenses, fees should be set at a level which would encouraged the efficient use of the resources allocated.

**A. Objectives**

The Community released the Licensing Directive in order to set the framework for general authorisations and individual licences.[362] In particular, the main objectives have been the following: first, authorisations of licensing regimes could provide the appropriate means to supervise access to market; second, licensing procedures should be harmonised; third, any conditions attached to licenses should be minimal, proportionate and transparent. The directive defined the generic term "authorisation", which comprised the two categories, "general authorisations" and "individual licences".

According to Article 2 of the directive, authorisation is "any permission setting out rights and obligations specific to the telecommunications sector and allowing undertakings to provide telecommunication services, and where applicable, to establish and/or operate telecommunications networks for the provision of

---

censing arrangements. The idea was for simplified procedures on the basis of harmonisation of the national conditions for a one-stop-shopping procedure. The next proposal of the Commission , COM (95) 545, was a proposal on a common framework for general authorisations and individual licenses in telecommunications.

[361] See Green Paper on Convergence, COM (97)623.

[362] Council Directive 97/13 OJ L No 117, 7.5.1997. General issues related to licensing were included in the Satellite Directive.

such services in the form of general authorisations or individual licenses". The difference between general authorisations and individual licences is that a general authorisation is regulated by a class licence or a general law, whereas individual licences need prior decision by the NRAs. (Article 2 Licensing Directive).[363]The Community gave a clear preference to the general authorisations, in order to promote and facilitate Community-wide provision of telecommunications services.[364]According to Article 3(3), individual licenses can be issued, where the applicant is given access to "scarce physical and other resources or is subject to particular obligations or enjoys particular rights".

In the general context of the licensing directive, the ETO proposed that as the directive applies to all categories of services, satellite services should not be considered a specific category, and, as a consequence, the licensing regime should be exactly the same as that of the telecommunications services.

In particular, the key elements of the directive were:

- the prohibition of any limitation in the number of new entrants, except for under limited circumstances and for a temporary period of time so as to ensure an efficient use of radio frequencies;[365]
- if Member States did not have a special regime for the provision of licenses, that is authorisations or individual licenses, they would not be obliged to introduce one;
- the priority given to general authorisations (characterised by the absence of prior approval by the NRA), as opposed to individual licences, was extended by the Council, and the principles set out by the directive have been very clear: national licensing regimes should adopt the least onerous systems possible and one of the means to achieve this goal is to resort as much as possible to general authorisations. The directive comprised a stipulation that "the provision of publicly available voice telephony services, the establishment and provision of public telecommunications networks, as well as other networks involving the use of radio frequencies, may be subject to individual licences;

---

363 As König has observed, this directive formulated the following types of procedures: on the one hand two harmonised authorisation procedures in the Member States for general and individual authorisations and one co-ordination procedure between the Member States, the one he called the «global procedure».See König/Zeiss, Lizenzierung von Kommunikationsystemen, EuZW, Heft 5/1999, p. 135.

364 See argument 7 of Directive 97/13 on Licensing.

365 See also Patrick Xavier, The licensing of telecommunications suppliers, Beyond the EU's directive, Telecommunications Policy, v.22, n.6, 1998, Kiessling, Blondeel, The EU regulatory framework in telecommunications, A critical analysis, Telecommunications Policy, vol. 22, 7, p.584.

- the definition of harmonised principles: the directive offered important guarantees to organisations applying for a telecommunications licence in the EU, especially for new entrants, by setting up time-limits and other procedural requirements and by establishing principles with which national licensing regimes will have to comply, particularly in regard to conditions which may be attached to licences. For instance, the directive forbids certain kinds of illegitimate licensing conditions, or requests that licence fees only cover the administrative costs and are proportionate to the work involved in issuing and monitoring the licence. However, these conditions should be minimal and transparent;
- in order to overcome the possible remaining barriers to the creation of an internal market for telecommunications networks and services, the directive also provided for a procedure allowing for the harmonisation of licensing conditions. Such a harmonisation procedure associated CEPT/ECTRA/ETO and CEPT/ERC/ERO in the technical preparation of harmonised conditions and should result in the Community decisions on creating rights for undertakings;
- the directive called for the creation of a licensing committee to supervise the licensing procedures. The Committee has a legal personality and is subject to the Community's rules;
- national administrations had to grant licenses in specific time frames, that is six weeks under normal circumstances, four months in specified cases, eight months in the case of comparative bidding procedures;[366]
- additionally, in the case of refusal of a license, which had to be justified, the doctrine of direct effect, as developed by the European Court of justice, with respect to Community Directives,[367] allowed companies which have been confronted with a refusal to grant licenses pursuant to Directive 97/13/EC to seek damages in national court proceedings in the case of either a lack of transposition or inadequate applications of Community Law provisions.[368] In addition

[366] Article 9-Procedures for the granting of individual licenses (...):
"2. Where a Member State intends to grant individual licenses:-it shall grant individual licenses through open, non-discriminatory and transparent procedures and, to this end, shall subject all applicants to the same procedures, unless there is an objective reason for differentiation, and –it shall set reasonable time limits; inter alia, it shall inform the applicant of its decision as soon as possible, but not more than six weeks after receiving the application. In the provisions adopted to implement this Directive, Member States may extend this time limit to up to four months in objectively justified cases which have been defined specifically in those provisions. (..)"

[367] ECJ 1974, 1337-Van Duyn: ECJ 1982, 70-Becker.

[368] Jointed cases Frankovitch (C-6/90 and 9/90), judgment of 19.01.1991; Jointed cases Brassiere du Pecheur and Factortame III (C-46/93 and 48/93), judgment of 05.03.1996; British Telecommunications (C-392/93), judgment of 26.03.1996.

they may ask the Commission to initiate an infringement procedure against the respective Member State.

## 6. Spectrum Green Paper[369]

### A. Background: The significance of spectrum as a scarce resource

Regulating spectrum has been difficult for the following two reasons. First, it is a scarce resource, difficult to allocate and assign, as the demand is much greater than the existing available frequencies.[370]Second, as spectrum assignment methods, due to their economic importance for national budget, have been an issue belonging to the national sovereignty of each state, it has been difficult to find a common formula for the harmonisation of assignment procedures. Regulators, however, agree that regulation on spectrum should correspond to market demands, even though the different mentalities lead to different confrontations, such as between the EU and the USA for instance, where the EC insists on co-frequency sharing while the US insists on the band segmentation approach.

Operational satellites of new systems require more frequencies, new frequencies. Spectrum availability has been a major preoccupation since the market in European is relatively small and land-based digital systems exist. This has made the compatibility between S PCS and land-based systems a necessity, since many buyers would demand interoperability between the two mobile systems. Furthermore, the Commission dealt with the question of which systems should be allowed to operate in Europe for the categories where there is spectrum scarcity in Europe. In the Commission a number of options for selection have been proposed, such as selection on the ITU basis, namely, the first come first served principle, selection of systems that have passed the ITU frequency co-ordination, on the basis of frequency auctions, on a "beauty contest“, or on natural selection process. [371] However, the elaboration of a spectrum licensing mechanism has to be coupled with the establishment of harmonised licensing regimes for the network and the provision of services for the S PCS systems.

---

[369] Green Paper on Radio Spectrum Policy in the context of European Community policies such as telecommunications, broadcasting, transport, and R&D, COM(98)596 final.

[370] It would be wise to distinguish the meanings of these three terms, as they are often confused: the term allocation refers to the entry of the new frequencies in the Table of Frequency Allocation in the Radio Regulations. According to the definition of the ITU it is "the entry in the table of frequencies of a given frequency for the purpose of its use by one or more terrestrial or space communication services." (ITU, RR, Article 1, No 17(1982). After frequencies have been allocated to services (and not to countries), they are allotted to countries. It is then the task of each government to assign the allocated frequencies. The term allotment refers to the entry of a designated frequency channel in an agreed plan to be adopted by a conference and used by the Administration for a station to use a radio frequency or a channel.

[371] COM (95)529, p.30.

Regulatory attempts both in the EU and internationally have tried to prevent conflicts of a global nature on spectrum.[372] As regards the efforts in the Community, these started in the eighties, with the introduction of the Directives on GSM and ERMES, and culminated in the Spectrum Green Paper, a policy document, which addresses all issues and tries to initiate a dialogue for spectrum policy harmonisation within the EU.

The path to trace a regulatory framework has involved various phases, such as planning and the use of radio spectrum, the harmonisation of radio spectrum allocation, the radio spectrum assignment, the radio equipment and standards, and the framework for radio spectrum co-ordination. Until the beginning of the eighties the situation in Europe as regards the frequency co-ordination and the administration of the orbit capacity was discouraging. Member States preferred to act individually and co-operation between the Member States and European institutions was very lax, if not inexistent. Furthermore, there was no institution at European level that could co-ordinate a common frequency policy between the EC and the Member States. CEPT, which was supposed to be the co-ordinating intergovernmental institution, was otherwise occupied defending the interests of the national telecom operators.

**B. The objectives of the paper**

The need for the creation of a common European Community spectrum policy in the context of the accomplishment of the single market and pan-European network has been the aim of the Spectrum Green Paper. The reasons for the Community's interest have been based on its major economic and political importance, and underline the need for the beginning of a new common policy for the radio spectrum.[373]

The paper's main objectives were to initiate a discussion on how, with the emergence of advanced technologies, one common spectrum policy could be shaped,

---

[372] Such disputes are usually not brought before the relevant body of the international regulatory organisation, which in this case, is the ITU's and its dispute settlement mechanism, due to the long time it takes to be mobilised. They are discussed, however, in informal meetings and procedures of negotiation. See Vineeta Shetty, Cover story: on a different wavelength, Communications international, 1 May 1999. Such points of conflict have been the following. The case of Iridium and Radio Astronomers transmissions, in the 1610-1660 Mhz band, where Iridium has been operating, the Radio Astronomers conducted experiments in the 1610.6 Mhz to 1613.8 Mhz band and these experiments were affected by Iridium's transmissions.

[373] Issues identified have been: (p.II) to secure radio spectrum availability, to ensure the appropriate balancing of commercial and public interest, to reflect its economic value, to provide legal certainty, to ensure open, transparent, objective and non-discriminatory procedures, to ensure the efficient use of frequencies, to ensure that relevant regulation encourages competition and technological innovation, and to support the development of economies of scale.

and to invite the Member States to co-operate in order to form a harmonised spectrum policy. Among the issues and questions which it posed, the paper dealt extensively with spectrum management (strategic planning, harmonisation of allocation, assignment, licensing, equipment and standards), including the following points:

- strategic planning in the light of the increased number of new systems as well as commercial and non-commercial market players;
- the allocation and harmonisation of its use to the radio spectrum, in the light of globalisation, which, on the one hand required international co-operation, though, on the other hand, countries attached different priorities;
- assignment of radio spectrum should be done under certain reasonable conditions and on the basis of market based approaches;[374]
- it would be important to monitor the use of radio equipment;
- institutional arrangements should be re-designed in order to be able to deal with the increased number of new players.[375]

---

[374] A distinction can be made between methods of frequency assignment based on administrative rules (i.e., the first-come, first-served method, and comparative-administrative methods), and those based on free market forces in which frequencies are assigned in accordance with the price mechanism (auction methods). At the beginning of the nineties except for market-oriented innovations in Australia and New Zealand, the most common economic approach to spectrum pricing taken by NRAs worldwide was on an administrative cost-covering basis. At the beginning it was the administrative methods that were being discussed. However, there has been a shift towards market based mechanisms. The more satellites started to be explored commercially, the more the value of the spectrum increased and the more governments had to ponder the commercial considerations of the technological aspect. The existing administrative procedures have serious disadvantages, such as inefficiency in controlling spectrum demand, very often pot-luck regarding openness and transparency in the assignment procedures, and difficulty in corresponding promptly to changes in both technology and market demand. These disadvantages have negative consequences for new interested parties, since the procedures often end up very either being very costly and under-utilising spectrum or creating difficulties for the market newcomers in comparison with the incumbent operators who are already well placed in the market. For frequency assignment methods, see Coopers & Lybrand, Study on analysis of new methods of frequency allocation in the Member States and comparative analysis of recent developments in this field, July 1993; NERA, Approaches to the Allocation and assignment of the radio spectrum resource, Executive Summary, Final Report for DG XIII, November 1997, p. iii. Also, S. Forge, The radio Spectrum, Telecommunications Policy, 20, 1, 1996, p. 68 ff., John Mc Millan, Why auction spectrum, Telecommunications policy, 19, n.3, 1995, p. 191-199, also Horst Siebert, (ed.) Wettbewerb und Regulierung in der telecommunications, 1995, p. 130-135; also on issues of economic frequency allocation: Jörn Kruse, Zugang zum Frequenzspektrum, in Ernst-Joachim Mestmäcker (ed.) Ein Symposium über Ordnungsprinzipien in Wirschaftsrecht der Telecommunikation und der elektronischen Medien (Kommunikation Ohne Monopole 2), Nomos Baden-Baden, 1995, p. 449 ff.

[375] Spectrum Green Paper, p. 2.

It is difficult to say how allocation, strategic planning, and monitoring should take place or what assignment methods should be used. Political mechanisms in such matters which are of great economic importance for the countries, have always had the first and the last decision-making power. Moreover, the important point is repeatedly mentioned in the Spectrum Green Paper, that the method that will be used must be based on a mechanism according to which the frequency allocations should take place according to objective criteria, and transparent, open procedures which guarantee accession on a non-discriminatory basis.[376]

The reasons for this behaviour by the Member States was, as explained in the previous chapters, mainly politico-economical: as S PCS projects were mainly US led and the European participation was limited, the Member States did not want to allow in this sector Community legislation which would mean the reduction of their state sovereignty and the facilitation of the establishment of US dominance. On the contrary, the GSM was a European project and European industries were very successfully involved in it.

### *V. Regulating S PCS*

#### 1. Technological and economic challenges to regulation

It was made clear that for the introduction of S PCS in the Community, the establishment of a technical framework not only for legal, but also economic and political decision making was necessary. For instance, the technical framework was related to the lack of spectrum and frequency allocation: S PCS demanded that all satellites and terminals operated at the same frequencies at a global level, so that every S PCS network could operate throughout the globe on the same frequency, without facing the problem of interference.

In this framework the Member States were invited to develop a Community policy and a co-ordinated position, in particular within the context of international organisations (ITU), through the Commission's co-ordination efforts between the Member States and to strengthen co-operation between ERC, ETSI, and ERO.[377]

---

[376] These criteria where established by the ONP Directive (Directive 98/10, OJ L 101), the Mobile Directive (Commission directive 96/2 amending Directive 90/388/EEC with regard to mobile and personal communications) and the Licensing Directive (Council Directive of 10 April 1997 on a common framework for general authorisations and individual licenses in the field of telecommunications). They mean that the criteria on which the selection is based are as far as possible unambiguous and cannot be interpreted in various ways. Furthermore, all interested parties may obtain information and observe the mechanisms. Moreover, the existence of equity is essential, if competition is to be facilitated.

[377] See Commission Communication COM (93) 171 and Scherer/Bartsch, Introduction, in Scherer (ed.), Telecommunication Laws in Europe, 4th Ed, 1998.

The idea was for the new technologies to combine the existing regimes instead of creating a totally new one, so as to, adapt to possible new needs. This was the case for S PCS, whose regulation in the European Community has been seen "as a process of evolution rather than of radical changes“.[378]The S PCS legal regime was based on a successful combination of existing satellite and mobile-based regulation adapted to the new S PCS technical specificities. In particular the Community tried to follow the same model that it followed with the mobile digital terrestrial systems, that is, first, the introduction of the systems and then discussion on the regulatory aspects. The existence of a regulatory framework for satellites and mobile terrestrial systems has been very successful, although the new technology, new actors and new market demands have raised regulatory issues which were not covered by the existing legislation. As a consequence, the introduction of the new idea gave rise to a number of considerations concerning issues such as frequency allocation, standardisation, industry participation and licensing.[379]

Regulating S PCS meant regulating the operation of the satellites in space, operation of the gateways and the other infrastructure, the provision of services, and the provision, circulation and use of the terminal equipment.

For example, some important points related to the S PCS regulation have been the following:

- gateways could be located in a different country to where the service would be provided. Permitting service provision would lie within the competence of the country where the end user would be. Gateways, being part of the satellite system infrastructure should comply with issues set by EC law such as safety or electromagnetic interference. Another issue would be the gateway operation; again there have been important issues such as the possibility of access. Each partner of the enterprise that operates the system might have had access, as well as every enterprise or governmental organisation that was authorised by contract to provide or receive services or facilities;
- owners of handset terminals should not be expected to pay custom duties to every country to which they travel and manufacturers should not suffer complicated type-approval regimes;

---

[378] See KPMG, on Satellite Personal Communications and their Consequences for European Telecommunications Trade and Industry, volume I, Report prepared for the European Commission, DG XIII, 1994.

[379] COM(97)217, Communications from the Commission to the Council, the European Parliament, the Economic and Social Committee and the Committee of the Regions on the further development of mobile and wireless communications. Challenges and Choices for the European union.

- processes of system licensing in the European Community should be non-discriminatory, open and fair and "technology transparent";[380]
- specifications on how to achieve interoperability between ground segment, handsets and the satellites;
- specification of tariff principles, interconnecting agreements in the case of providing services to a third country;
- for the operation of the ground segment, S PCS operators would need more licences, as the one for space segment would not cover their right to set up earth stations, operate gateways and connect to the public switched telephone network. The earth stations need new frequencies, different from those that use the satellites to transmit the signals as well as licenses to install the earth ground infrastructure;
- For the MES, efforts would aim to create of a simple one stop shopping licensing regime and the simplification of type approval-procedures so that they are without any further formalities. Moreover, for the provision of services the service providers need to be licensed;

## 2. Decision 710/97 on S PCS

### A. The decision

Decision 710/97 of the Council was released in order to speed up the introduction of compatible satellite personal communications services in the Community through the harmonisation of frequency bands, the removal of remaining barriers to the free movement of terminal equipment and usage conditions attached to general authorisations, and the simplification of licensing procedures. This decision constituted, in particular, the first Council decision on S PCS, showing the importance of these communications and underlining that a co-ordinated action would be urgent and necessary: the S PCS sector would be particularly complex, since it dealt with global systems that intended to serve users on the move, within or outside the EU, thus implying third country action. [381]

The decision of the European Parliament and of the Council gave an important role to the CEPT Administrations, containing a brief for the organisation to harmonise by September 1997 the frequencies and authorisation conditions relating to S PCS, based on mandates issued by the European Commission.[382]

---

[380] The term "technology transparent" signifies that regulation should not concentrate much on the technology used, but more on the result of the technology used, that is the quality of the services offered.

[381] Scherer/Bartsch, Introduction, in Scherer (ed.), Telecommunications Law in Europe, 4th Edition, 1998.

[382] See Articles 2 and 3 of the Decision

As regards CEPT/ERC responsibilities related to the field of the EU frequency and licensing regime, this decision came in addition to the Commission's Communications of 1993, in which the Commission recognised CEPT/ERC as the institutional bodies responsible for harmonisation's policies.[383]Furthermore the Council pointed out the possible necessity of the intervention of standardisation bodies such as ETSI, CEN and CELENEC to prepare the necessary standards. Moreover, the following points were underlined:

- the right of each country to decide whether or not, and how, it would participate in satellite personal communications systems as well as to determine the conditions and terms of participation for the uplinks, as this would be a sovereign right which operators would have to respect. Operators had to follow the procedural requirements that each administration asked for. However, the global nature of the services would made the establishment of a one-stop-shopping procedure necessary, under certain conditions and would also necessitate the harmonisation of national policies of the administration.
- the Council authorised the Commission to start multilateral and bilateral negotiations concerning the S PCS on the basis of specific mandates from the Council, in order to enable the effective access of Community operators. Furthermore, the Commission may give mandates to CEPT/ECTRA or CEPT /ERC to harmonise frequency use and the conditions attached to general authorisations for satellite personal communications services.

Responsible for the implementation of the Decision, namely, for the co-ordination of national authorisations, the Co-operation of the European Commission with CEPT/ERC/ECTRA and the elaboration of a one-stop-shopping procedure would be the Licensing Committee, set up by the Licensing Directive. The Licensing Committee has a legal personality and comprises of representatives from the national administration of the Member States. In particular, the Decision established a quite complicated procedure, in which the Commission is almost forced to co-operate with the Licensing Committee.[384]

---

[383] COM(93)382: The Commission's New Approach has constituted an instrument of policy and strategy planning between the Commission, CEPT/ERC and the Member States, setting out principles governing Union action and Commission proposals.

[384] Article 6 describes the procedure according to which the Commission submits to the Committee a draft of the measures to be taken. The Committee delivers its opinion to the draft within a time limit which the chairman may lay down in Art. 148.2 of the Treaty in the case of decisions which the Council is required to adopt on a proposal from the Commission. If the measures taken by the Commission, which are directly applicable are not in accordance with the Committee, they shall be communicated to the Council. In this case the Commission either defers application of the measures, or the Council acting by qualified majority may take a different decision.

In brief, the decision set a legal framework, comprising of regulatory bodies and committees of expertise: the European Commission, ERC, ECTRA, as well as the Licensing Committee.[385]The law-generation process has worked with the mechanism of the mandates, according to which the European Commission, in close co-operation with the Licensing Committee, gave the mandates to ERC and ECTRA for the ratification of decisions that would on an intergovernmental basis harmonise policies, in particular on spectrum or licensing procedures. This mechanism has functioned as a safety net for the market players in order to secure that there would be enough spectrum in the market. Furthermore, according to the mandate mechanism, the political weight of decision-making went to the Member States, as ERC was an intergovernmental body. The mandates were discussed in meetings with the licensing Committee where the Commission, ERC, ECTRA and national delegations participated. If adopted, ERC and ECTRA would work on the mandates and release decisions relating to frequencies or licensing procedures and the national administrations of the Member States would be informed. As a consequence, the Commission has given a series of mandates demanding high specification concerning the harmonised use of frequencies, the harmonised conditions attached to general authorisations, the free movement of terminal equipment in the Community, harmonised authorisation procedures for Big LEOs,[386] and Giga LEO frequency introduction, which "may be considered S PCS".[387]These mandates have been given in order to facilitate the entry into the European market of systems, such as Iridium, Globalstar, (Big Leo), Orbcomm, E-Sat (Little Leo) as well as Teledesic, and Skybridge (Giga Leo).

**B. S PCS decision - Licensing Directive**

The S PCS decision includes a specific regulation on S PCS, whereas the Licensing Directive concerned licensing for the whole sector of telecommunications; the general rule applicable in all telecom cases. The Community considered licensing a top priority and, in particular, in the S PCS case, the licensing of the global satellite systems had a particular importance very much related to US regulation. The Community considered S PCS specific regulation of licensing a

---

[385] See Articles 2, 3, 4, 5, 6, 7 of the Decision.

[386] See The EU Satellite Action Plan, (SEC) 1999, p. 13.

[387] This was the wording of the mandate given by the Commission to the ERC for S PCS above 3 Ghz, which may be considered as an S PCS. According to the S PCS decision, systems either below 3 Ghz or between 1-3 Ghz are characterised as S PCS. With this mandate, the Commission includes satellite broadband systems above 3 Ghz in the S PCS category. With this mandate, the Commission wanted to catch up with recent technological developments, with the existing legal framework, until the proper one was found. The mandate was given by the Commission to the ERC and announced in the Licensing Committee of 13 February 1999.

necessity, since S PCS, such as Iridium, Globalstar and Orbcomm wanted to start operation. In the Community there was no regulatory frame, whereas, in the US, the rules for licensing had already been set in 1993.[388]Co-ordinated licensing regimes would signify the saving of time and money for the S PCS operators and would be a good reason for the encouragement for the European market players involved in S PCS, such as Alcatel, Alenia, and Vebacom.

As the S PCS Decision mentioned (argu. 6), due to the fact that the licensing directive has been a general directive, that these matters were urgent and that no such authorisation measures have been adopted at Community level, the action taken on S PCS was required for their more rapid introduction. An important part of the[389]decision has been Article 3, which defined the specific mandate mechanism, and mandated the ERC for frequency co-ordination and ECTRA for general authorisations. As regards the one-stop-shopping procedures the framework has been the same. CEPT/ERC and CEPT/ECTRA both have a legal binding framework: they have to issue soft law documents under the mandates that they receive.

The decision 710/97 was extended through a new decision of the European Parliament and the Council until 31 December 2003.[390]

### 3. CEPT/ERC

The mandates, which were given by the European Commission to CEPT were successful, as it was much easier for the Member States to deal with non-legally binding decisions made by intergovernmental institutions on issues such as spectrum assignment and authorisations. With the mechanism, where regulatory non-legally binding competencies were transferred to ERC/ECTRA[391]all players were satisfied, that is, the Member States, the European Commission and the industrialists: the Member States because they did not see their national sovereignty being threatened and the industrialists because of the flexibility caused by

---

[388] Syed Kammal, as above.

[389] In its annex the Licensing Directive refers in to public security and to the conditions which may be attached to authorisations.

[390] Decision no 1215/2000/EC of the European Parliament and of the Council of 16 May 2000 extending Decision No 710/97/EC on a coordinated authorisation approach in the field of satellite personal communications services in the community, OJ L 139, 10/06/2000, p. 1.

[391] A series of documents, which deal with this issue, enacted from 1992 until 1997 exist. The first one was the Council Resolution on the implementation in the Community of the European Radio communications Committee, (OJ C No 318/01 of 19.12.1992) which has constituted the first milestone of the Community's policy on co-operation with ERC. The second important document was the "New Approach" [COM(93)382]. Two years later, the Commission released a new Communications of the Commission to the Council and the European Parliament on the evaluation of the CEPT/ERC Decision mechanism [COM (95)85].

a relative loose framework compared to that of the directives. As a consequence, ERC and ECTRA adopted four decisions, which played a major role in the effort of the Community to harmonise spectrum assignment and licensing as every participant showed willingness to co-operate.

**A. BIG LEOs**

a. ECTRA/DEC(97)02 [392]

This decision constituted a significant contribution by the ECTRA to the efforts of the Commission for harmonised policies as regards the licensing regime in the field of Satellite Personal Communication Services in Europe. It established a relatively complicated procedure involving administrative and regulatory bodies, and industrialists. The purpose of this complex procedure was to assure that the applicant indeed wanted to make the system operational and that the services provided would have the quality that the customers would expect. The decision underlined the fact that the granting of authorisations for the operation of S PCS systems belonged to the states, that is the NRAs, which may also require individual licenses. The decision specified the term "authorisation", (which, in the Licensing Directive, obtained a very broad significance, since it referred to the whole telecommunications sector), in the S PCS sector. In this framework the term meant legal or administrative measures regarding public telecommunications networks or services that grant access to frequency spectrum for use by the Mobile Earth Stations of S PCS systems, licensing and free circulation of MES within CEPT, also the service provision of S PCS, installation of gateway stations, and interconnection to public networks.

b. ERC/DEC(97)03 393

The decision referred to the harmonised use of spectrum for the satellite personal communication services operating within the bands allocated by the WARC 1992.[394] This decision constituted an effort to provide a harmonised policy as far as the use of the designated bands for the S PCS in the CEPT Member States was concerned. The previous attempts at EC level, taken by the Commission and the Council, were aimed at harmonising frequency bands for the introduction of GSM, DECT and ERMES. Harmonised frequency assignment is considered to be necessary in Europe in order to facilitate the efficient use of spectrum. Furthermore, the decision reconfirmed that in order to allow MES of

---

[392] The ECTRA decision of 3.7.97 on harmonisation of authorisation conditions and co-ordination of procedures in the filed of S PCS in Europe, operating within the 1.6-2.4 and 1.9/2.1Ghz bands.

[393] ERC/DEC/(97)03 of 30 June 1997 on harmonisation of authorisation conditions and co-ordination of procedures in the filed of S-PCS in Europe, operating in the frequency bands 1.6/2.4 and 1.9/2.1GHz.

[394] 1610-1626.5 MHz, 2483,5-2500 MHz, 1980-2010 MHz and 2170-2200 MHz.

an S PCS system to operate on these frequency bands, it would be important that S PCS had already been granted the permission to operate in the relevant band by the ITU. With this decision, ERC was called to play the role of the "mediator" between the ITU and the EU and to push the Member States in order to speed up the procedures for the elaboration of new frequency assignment tables, according to the allocations made by the ITU. The faster the Member States made the tables, the quicker the systems could be introduced in the Community.

c. ERC/DEC(97)04
This decision was taken to deal with transitional arrangements for the fixed service and Mobile satellite service in the 1.9/2.1 band in order to facilitate the harmonised introduction and development of S-PCS. When new systems are to be introduced and new frequencies are allocated and assigned, the old frequencies have to be relocated and new arrangements have to be made. With this decision, the ERC aimed to facilitate the transitional arrangements and keep all players satisfied.

d. ERC/DEC (97)05[395]
In the framework of the harmonised policy on spectrum use and on licensing procedures, that is, harmonised authorisation procedures, co-operation between the NRAs, establishment of one-stop-shopping procedures and promotion of the policy of general authorisations instead of individual licenses, the decision (97)05 facilitated the licensing regime for the Mobile Earth Stations of the S PCS. It was based on the regime elaborated by the licensing directive. The decision was also elaborated within the framework of a formation of a free circulation for telecommunications equipment, like the older decision ERC/DEC /(95)02 for the free circulation of radio telecommunications equipment.[396]This policy was facilitated by the Satellite Directive 94/46 which equalised satellite ground stations with telecommunications equipment. The decision on the one hand recognised the right of the NRAs to require several types of individual licenses for network operators, services providers, gateways and mobile earth stations. On the other hand, however, because of the mobility of the network and of the global character of the system, the decision suggested that Mobile Earth Stations should circulate free in the Member States of CEPT, without requiring individual licenses by the Member States, provided they meet three further requirements. First, the Mobile Earth Station operates in the identified frequency

[395] ERC/DEC/(97)05, ERC Decision of 30 June 1997 of free circulation, use and licensing of Mobiel Earth Stations of Satellite Personal Communications Services operating in the bands 1610-1626.5 Mhz, 2483.5-2500, 1980-2100, 2170-2200 Mhz within the CEPT.

[396] In the decision, the categories of type-approved radio equipment were considered to be: GSM mobile phones, DECT mobile equipment, Omnitracs terminals for the Euteltrack system, Inmarsat-C terminals, Inmarsat-M terminals and PR-27 mobile stations.

band by the ERC/DEC/(97)03 decision. Second, the S PCS system has been granted an authorisation and third, the Mobile Earth Station fulfils the further requirements of the country.

e. Procedures for granting of an authorisation to the applicant
Moreover, the decision described the authorisation procedures, which were initiated if an entity submitted an application to the responsible NRA. In particular, Annex 1 of S PCS scheme pointed out that "an entity (applicant), acting within the national law of an administration, seeking authorisation within that administration or a number of administrations, submits an application to the administration". The application should include, first, information about the applicant and its project, referring also to the ITU's advanced publication and to how the applicant would fulfil the milestones. The application would then be examined by the administration, and its compliance with the ERC/DEC/(97)03 and ERC/DEC/(97)05 would also be considered. If the requirements of the national law, ECTRA/DEC/(97)02, ERC/DEC(97)03 and ERC/DEC(97)05 were fulfilled, then the administration would forward the application to the MRC. The MRC again examined if the application fulfilled all the requirements. If the result of the examination was positive, then the MRC proceeded and examined whether the application fulfilled the eight milestone criteria.[397]Only when all cri-

[397] According to ERC/DEC(97)03, these are: 1. Submission of ITU Advance Publication and Co-ordination documents. The satellite network operator had to provide clear evidence that the administration responsible for an S-PCS system had submitted ITU RR Appendix 4 'Advance Publication Information to Be Furnished for a Satellite Network' and ITU RR Appendix 3 'Notices Relating to Space Radiocommunications and Radio Astronomy Stations'. 2. Satellite manufacturing. The satellite network operator had to provide clear evidence of a binding agreement for the manufacture of its satellites. 3. Completion of the Critical Design Review. This is the stage in the spacecraft implementation process at which the design and development phase ends and the manufacturing phase starts.4. Satellite launch agreement. The satellite network operator had to provide clear evidence of a binding agreement to launch the minimum number of satellites required to provide a continuous service within the CEPT.5. Gateway Earth Stations. The satellite network operator should provide clear evidence of a binding agreement for the construction and installation of Gateway Earth Stations that would be used to provide S-PCS within the CEPT countries. 6. Launch of satellites: a. The satellite network operator should make available for inspection by the MRC documents confirming the first successful satellite launch and in-orbit deployment. b. The satellite network operator of an NGOSS system should also provide periodic evidence of subsequent launches and successful in-orbit deployment of satellites in the constellation. 7. Frequency co-ordination. The satellite network operator should submit to the MRC documents relating to the successful frequency co-ordination of the system pursuant to the relevant provisions of the Radio Regulations. However, a system, which demonstrated compliance with milestones 1 to 6 inclusive was not obliged to demonstrate at this stage completion of successful frequency co-ordination with those S-PCS systems, which failed to comply adequately and reasonably with milestones 1 to 6 inclusive. 8. Provision of satellite service within CEPT. Before 1 Janu-

teria were fulfilled, could have the national administration granted a license of operation. With the granting of an authorisation, the Member States could also attach certain conditions, which had to be based on equality, transparency and proportionality and should comply with the EC competition rules for EC Member States. General conditions could be attached to all authorisations. If all milestones were successfully fulfilled, the national administration assisted by the MRC, gave the authorisation to the operator. The milestones were demanding and the whole process for obtaining a license rather strict. However, it has been the only way to ensure quality in personal communications. As soon as an applicant system provided evidence of the first successful satellite launch and in-orbit deployment, administrations were supposed to proceed with frequency assignments and the authorisation procedures.[398]

**B. Little Leo Regulation**

The first regulatory attempts had focused on the creation of a regulatory framework for the Big LEOs, as the first players were interested in Big Leo systems. As a consequence, a long administrative procedure requiring compliance with milestone criteria and constant co-operation with the National administrations and the MRC had been established for the Big LEOs, while for the Little LEOs there was nothing. However, after Orbcomm became operational and various other Little Leo systems, such as Safir, and E-Sat showed interest in starting operation, the creation of a regulatory framework also for these kind of constellations became a necessity as well. After a mandate had been given by the Commission to ERC, the latter released a recent decision (March 99) in order to coordinate the national administrations introducing regimes for the Little LEOs.[399] This decision constitutes major regulatory and administrative help to S PCS regulation in Europe.

The decision also constituted a step towards the introduction of S PCS <1 GHz in the CEPT countries, providing a common approach for the establishment of

---

ary 2001 the satellite network operator had to notify the MRC that it had launched, and had available for the provision of services the number of satellites it had previously identified under milestone 4, as necessary to provide continuous commercial service, and that it would be providing commercial service within the CEPT using the frequency bands identified in the relevant decision.

[398] In particular, ERC/DEC (97)03 specified that .."for an S PCS system which meets all milestones up to and including milestone 6[..], and which becomes operational and ready to provide commercial services within the CEPT prior to 1 January 2001, its MES may operate on an provisional basis, subject to national authorisation in the relevant administration which may be conditional to the outcome of frequency co-ordinations with other services in that country."

[399] ERC/DEC/(99)06 on Licensing of Little LEOs.

procedures based on a case-by-case analysis.[400] The main points for consideration were the identification of frequency bands of operation as well as the elaboration of evaluation procedures. The important parts of the decisions were the 3 annexes. In these annexes the following have been defined: first, the system notified the ITU and at least one CEPT country (Annex 1); second, systems that satisfied all the prescribed conditions of the decision and have made a successful launch of the first satellite are, therefore, to be considered for introduction (Annex 2), third, the compliance to milestone criteria (Annex 3).[401]

Furthermore, the decision foresaw close co-operation between the ERC, whose competence are the S PCS systems below 1Ghz, and the national administrations and puts in question whether or not the milestone evaluation process would be under the responsibility of the MRC, which is responsible for the Big LEOs or of another body. Moreover, like all other ERC decisions on S PCS, the systems would have to prove operation within the frequencies assigned until the 1st of January 2001, after which the decision would be reviewed and, depending on spectrum requests and competitiveness in the sector, spectrum assignments will become definitive. The decision introduces the Milestone Criteria, which all candidate systems have to fulfil. However, for the introduction of a system in the CEPT countries it is only important to fulfil the first six.[402]

By introducing the harmonisation of the criteria within CEPT administrations, this decision attempted to co-ordinate spectrum policies also related to the spectrum needed for the S PCS below 1 GHz, that is, spectrum assignments should take place within the allocated bands, taking all specificities mentioned in the decision.[403]

---

[400] The points for a common approach, as mentioned in the decision have been the following: first, to identify the bands below 1 GHz for the MES of the Little Leo systems. Second, to establish technical and operational constraints, related to the specific type of S PCS <1 GHz for the frequency use and compatibility with terrestrial systems. Third, to promote competitiveness and elaborate criteria for future Little LEOs to be brought into operation in the same frequency band, fourth, to adopt a milestone evaluation process in order to avoid paper satellite systems.

[401] The decision provides that one of the systems listed in Annex 1 will be listed in Annex 2 upon request of at least one of the CEPT administrations, provided that a) this specific S PCS <1 Ghz has met milestones 1 to 6, b) all necessary interservice and intraservice compatibility studies for this S PCS<1 GHz system have been successfully completed and agreed by the ERC and c) the required operational constraints have been approved by the ERC.

[402] As in the Big LEOs, the criteria are related to submission to ITU of advance publication, satellite manufacturing, etc.

[403] Bands assigned are 137-137.025 MHz, 137.175-137.825 MHz and 400.15-401 MHz (space-to-earth) and 148-149.9 MHz, 149.9-150.05 MHz, 399.9-400.05 MHz and 406-406.1 MHz (earth-to space) allocated on a primary basis. Bands allocated on a secondary basis are

Moreover, as regards the individual licensing of the Mobile Earth Stations (Terminals), other recent ERC decision on the MES of S PCS <1GHZ have established a regime where S PCS <1 GHz terminals are exempted from individual licensing. [404]

As seen, these two decisions follow the example of the ERC decisions on the Big LEOs. The decision on the milestone criteria is demanding as the systems are global and administrations have to make sure that customers will enjoy quality service.

***VI. Summary***

The introduction of the S PCS in the Community at the beginning of the 1990s was only a goal; it was not driven by any legal principles or obligations. It was only in the phase of the market entrance of the S PCS systems that a legal obligation was set. In the majority of the countries, both European and non-European, the lack of a legal regime as well as the lack of co-ordination among the authorities was a major obstacle. Through the Council, the European Community decided to take the initiative and enact a framework for co-ordination.

In this chapter, mainly the structural elements of law have been analysed. Law, as a set of rules and procedures has been involved in all levels of interaction. Emphasis has been given to the international and European level playing field, where law has had in principle a flexible, general and de-detailed form, the degree of flexibility being higher at the international level than the European one. In contrast, at national level, rules have become more concrete. For the national administrations of the EC Member States the elaboration of legislation in the spirit of liberalisation has become a very difficult task, since legislation must regulate advanced technology, in the proper form of the country's specificities.

At international level ITU has been the collective mechanism of international-decision making. In the framework of the ITU, the issues concerning satellite communications started to be regulated from the end of the 1940s, when the ITU became an official UN body.[405]The It's Radio Regulations have provided the regulatory context. Regulation has been covering, in general, all issues concerning radio waves, and, in particular, issues such as frequency and orbit assignment, and harmful interference, as well as the circulation and licensing of the Mobile Earth Stations for the non geostationary mobile satellite systems. The

137.025-137.175 MHz, 137.825-138 MHz and 387-390 MHz (space to earth) and 312-315 MHz (Earth to Space).

[404] See the ERC Decision of March 1999 on free Circulation, Use and Exemption from individual Licensing of Mobile Earth Stations of S PCS <1GHz systems, ERC/DEC/(99)05.

[405] It was in 1947 during the Atlanta Conference.

mechanism of non-legally binding decisions has had positive and negative results. On the one hand, it has attracted many members and consequently general regulations on important issues such as harmful interference could be set. On the other hand, it has led to many conflicts especially between developed and developing countries.[406]One could conclude that the fragile balance between political actors attracted the interest of big enterprises and strategic alliances, which found the chance of penetrating the ITU playing field thus influencing the negotiations. For instance, during WARC 92, big enterprises such as Motorola, Alcatel and Ericsson got involved in order to influence national delegations and vote for frequency allocation for MSS. Moreover, during WRC 95, Teledesic representatives, with the help of the US delegation, managed to get the 500 Mhz band for operation.

The form of the international non-legally binding law has also been adopted at the European level, where two different, but co-ordinated sub-sites are to be found, on the one hand, the European Community and on the other the CEPT with its subcommittees. In the European Community, the European Commission, having the support of the Council, has, from the beginning of the nineties started to formulate an active policy first on satellites and then on S PCS. As regards CEPT, at the beginning of the nineties, it was transformed from a body of national network operators with technical responsibilities to a body responsible for the harmonisation of frequency policies in Europe. The ERC and the Commission have done a very important task as regards S PCS regulation, which the majority of the European Community Member States have difficulties in following. For example, none of the Member States has introduced separate legislation for S PCS, although all had introduced new licensing regimes.

The legislation has been presented separating the framework for general satellite communications and for the S PCS, since S PCS legislation is a legislation based on the existing regulation for satellite communications plus the new regulation which is mainly based on the soft law model, which is progressively being introduced in order to cover the specificities of this field. As a consequence, the S PCS legal framework is mainly based on non-legally binding rules, whereas the majority of the satellite communications law is hard law. For example, satellite communications have been regulated with important directives, such as the Satellite Directive (94/46), the Satellite Equipment Directive (93/17, the Directive

[406] See, for instance, the Bogota Declaration, the conflicts on orbital slot assignment. See for more, Martin A.Rothblatt, The impact of International Satellite Communications Law upon Access to the Geostationary Orbit and the Electromagnetic Spectrum, 16 Tx., Intern'L L. J., 1980, 207 ff. and Paris Arnopoulos, The international Politics of the orbit spectrum, Annals of Air and Space Law, v.7, 1982, p.215-240.

on Mutual Recognition of Conformity for (Satellite) Terminal Equipment (98/13) and the Licensing Directive (97/13). The majority of the S PCS legal framework consists of ERC decisions, which are representative examples of the soft law. [407]

The recent legislation shows that the Community has focused on certain fields, which are: mutual recognition and the conformity of assessments of the terminal equipment, licensing and issues of competition, and harmonisation of spectrum assignment which are related to the market for S PCS.

This has been the legal framework provided; however businesses have spectacularly failed. The strategies that lead to failure will be thoroughly analysed in the next chapter.

[407] For instance, the Decision of the Council 710/97 on a common authorisation approach in the field of S PCS systems is a hard law document. Furthermore, the Directive 98/13 on mutual recognition refers to S PCS systems indirectly.

# Chapter VI
# The strategies of the Strategic (S PCS) Alliances

## *I. Introduction*

Entities engaged in economic activities are interested in acquiring more resources and becoming more profitable. To this end, they develop a certain behaviour in the market on the basis of the power they have and their objectives, be they tracing new profitable market segments and expanding, investing in resources and R&D or searching for new partners to team up with.

The success or failure of an economic entity in the market may depend on many factors that can be related to the technology developed (i.e, degree of innovation, applicability), the financial support that will be found or the legal and administrative requirements (i.e., high barriers to entry, which may include very high spectrum fees) that will have to be applied. However, the success or failure of the product or of the services is in many cases decided by the market itself as it is related to factors which, if not taken under serious consideration, like in the case of Iridium, may lead to failure. In particular, issues related to marketing plans, strategies of entry of the alliances, market size, customer needs and service segments need strategic considerations in order to make sure that the billions of dollars invested will guarantee profit.

The third chapter of this study has presented the S PCS systems some of which became operational, others, however, despite making an enthusiastic appearance, were withdrawn or sold, before even entering the market actively. The strategies followed by the companies did not prove to be enough to attract a mass market. The Big and Little Leo systems that became operational found themselves with few thousands of customers and either became bunkrupt and were sold for a few thousands of dolars or were faced with severe financial difficulties and remained stagnant. The analysis of the strategies of the various economic entities as well as the reasons that lead the businesses to fail is the point of focus of this chapter.

## *II. Industrialists involved in the S PCS*

Until the mid 1990s the satellite industry was considered to be a stable and predictable sector, in which resource exchange and patterns of interaction were well organised and relatively stable.[408]However, in recent years there have been many transformations, as the industry has been undergoing a process of consolidation and co-operation. The rapid technological growth of the satellite commu-

---

[408] See Willem Hulsink, Privatisation and Liberalisation in European Telecommunications, 1992, p. 2.

nications in the telecommunications sector and the ever increasing involvement of the manufacturers in the satellite communications business has given the push for more industrialists to get involved in the sector. Since the late 1980s, many firms from the computing, electronics and telecommunications industry have started getting involved in various interfirm collaboration agreements. The complexity in the chains of production, supply and marketing have made satellite markets part of the telecomms network industry, sketching out the multiplicity of actors involved in the production and supply chain, and illustrating in this way various vertically integrated levels interrelated with the scope of the delivery of products and services to the final consumer via a "network infrastructure".[409]

The increased number of strategic alliances formed during recent years in the satellite sector has made clear that the only way to survive in the market and face up the enormous amount of resources required in the S PCS sector is through partnerships, as one single company and even successful multinationals would find it too difficult to compete both in terms of resources and investment. Alenia, Hughes, Motorola, Matra Marconi, Mitsubishi, have all chosen to enter the S PCS market through alliances, whereas in the mobile cellular land-based market the situation has been different: in this market firms like Nokia, Motorola and Ericsson have managed on their own to dominate the market, on the basis of each firm's resources and the need for partnership has not up to now been necessary.

In particular, industrialists developed the strategy of multiple participation. This strategy meant that they were not only engaged in one kind of the production or supply chain, or even in one kind of technology, but on the contrary, they expanded as much as they could, by identifying new technological innovations and targeting new market segments. As a consequence, they got involved in all sectors, where profits could be made. For instance, Loral has not only been one of the most powerful satellite manufacturers but has also been an investor in satellite projects. Furthermore, Qualcomm has not only been an investor in the S PCS system Globalstar, but has also been one of the most successful microchip manufacturers.

---

[409] See Center for Economic and Policy Research, Europe's network industries: conflicting priorities, Monitoring European Deregulation, 1998, p. 5. The term consists of three key parts, which are the following: first, the core products; second, network infrastructure; and third, the customer supply and service provision, or else the upstream production-infrastructure and the downstream service provision. In the network industries the core products and services are supplied to the network infrastructure by firms in the upstream. Then, providers that lay down stream, provide the final consumers with the services.

These and many other examples show clearly that industrialists involved in the S PCS sector have developed many expansion strategies, the recent ones being expansion in potential big markets, like the Eastern Europe, China and Latin America. With their technological and financial power they are able to manipulate governments and play a decisive role in the decision-making processes.[410] Their role in the industrial supply chain has determined their strategy, be they individual entities or acting in a consortium.[411]

### *III. S PCS - Big LEOs*

#### 1. Iridium

Iridium became operational in November 1998, two months later than expected as the handsets were delivered later than the agreed time and this delay resulted in the cancellation of many of the customer orders.[412]Before becoming operational, it expected to have 5 million customers around the world by 2002, of which about 1 million would be in North America.[413]However, on 31 December 1998 -two months after the beginning of the operation of Iridium- there were only 3000 activated subscribers, providing a revenue of $186.000, which reflected Iridium's five week delay in launching its world-wide commercial service and a shortage of Iridium telephones and pagers during the first two months of operation.[414]This figure was 6000 or 7000 less than analysts had expected it to be.[415]Iridium hoped to increase the number of worldwide telephone subscribers by the end of 1999.[416]However, reality has been hard, as figures have shown a much lower number of customers.[417]The system mainly focused on the inter-

---

[410] Their strategy has been to enter the market from the lower end, and from the most regulated side of the market, "in order to get a strong foothold in preparation for the next wave of large-scale infrastructure". See Annemarie Roobeek & Jeroen Broeders, Telecommunications global restructuring at full speed, in H.W de Jong: The structure of the European industry, Kluwer Academy Publishers 1993, p.291.

[411] Individual entities as strategic actors have been presented in Chapter III.

[412] For instance, from the beginning of operation at the end of 1998 Iridium waited for 70.000 handheld terminals to be delivered by Kyocera. However they were not delivered. This had a negative impact on the orders, as Iridium had to cancel many of them. Theresa Foley, Iridium and Globalstar hit delays, Communications Week International, 21 September 1998.

[413] Andrea Brooks, Iridium Settles Freedom to roam, Bloomberg News 25 February 1998.

[414] The initial forecasts of Iridium for the beginning of 1999, taking into consideration the delays, were for up to 25000 customers. See Iridium LLC, Iridium Reports Fourth Quarter Results, 25 January 1999. Also Guy Daniel, Pie in the Sky?, Communications International, 1/1/99.

[415] Mobile Communications, February 4th,1999.

[416] To 500.000 and 600.000

[417] However, experts estimated that Iridium would not be able to add more than 250,000 subscribers to its network during 1999. See Research Alert in biz.yahoo.com/rf/990203/ui/html; also GMPCS: Now the talking can begin, PNE, October 1998.

national business travellers, who needed to be able to make calls in areas where there was not a wireless telephone service.[418]

Due to financial difficulties Iridium has continuously borrowed from the bank, and its financial support started drying up.[419]As a result, at the beginning of June 1999, bankers gave Iridium a short period of time to pay back a debt of $800 million. After two months, Iridium filed for Chapter 11 for bankruptcy protection. In order to save the situation, Iridium started cutting stuff, dropping prices and reorganising the market structure.[420]The 55.000 subscribers in the year 2000 where not even enough to pay interest on the project's start up costs; Iridium would need at least one million subscribers to even get close to breaking even.[421] The US Court had followed the course of Iridium and gave the bondholders the right to sue Motorola. After the filing against bankruptcy, Teledesic became interested in Iridium, however, it was stated that no more funding would be given to the project. The provision of services had stopped, and Motorola was going to pull satellites from their orbit, however, its assets were sold at an auction and the 5 billion project was sold for some millions of dollars to Iridium LLC in 2001.[422]Since then, there has been a restructuring of the project: smaller infrastructure (one gateway instead of twelve), cheaper call prices ($1.50) and handsets ($1500) but the number of customers has not increased as expected.

Iridium was too certain of the market; it followed the marketing and distribution methods that costed more than expected, and as a consequence, the capital needed ended up to be much higher then initially estimated.[423]Iridium did not form a proper customer oriented basis and, in addition, has made some serious

---

[418] Iridium's primary market has been the international business traveller market (45% to 55% of expected initial customers for voice services), users in rural and remote areas (15% to 25%), governments users (15% to 25%), business communications (5% to 15%), while aeronautical customers would account for less than 1%. International travellers were also the primary market for paging services, which however Iridium gave up after the first months of operation.

[419] See Theresa Foley, Impact felt as highflier Iridium falls to earth, Communications Week International, 24/5/99.

[420] In particular it has dropped the prices of the calls from $ 7 a minute to $1,89 and $3,99 as well as the prices of handsets from $4000 to $3000, also, see, No white knight for Iridium, in www.totaltele.com/secre/view.asp?ArticleID=26458&pub=tt&categoryid=0.

[421] See, The return of Iridium , in http://www/forbes.com/2001/11/30/1130tentech.html.

[422] See No white knight for Iridium, in www.totaltele.com/secre/view?asp?ArticleID=26458&pub=tt&categoryid=0.

[423] In particular, Iridium's marketing strategy has been characterized as one of "the most mismanaged pieces of corporate planning ever seen". See Public Network Europe, October 1999, p. 65.

mistakes, which could, if avoided, have guaranteed a success.[424]First, the gateway operators chosen were inexperienced, as some were part of consortia with no experience in telecommunications. Second, the service providers that constituted the key interface with the customers were untrained. Third, there was a flawed business concept. The market of international travellers -main market target of Iridium- has proved to be limited and the other segments, like gas industries have been too tiny to support a mass market. Moreover, these sectors are already very successfully covered by the services that INMARSAT provided. Fourth, despite the design of the system that lasted ten years, the launch took place too soon. Iridium decided to launch, despite the technical problems that the systems were experiencing, something which made customers unhappy. Fifth, the prices were very high and the GSM prices that were getting even lower made Iridium unattractive to customers. The shape of the handsets, compared to the lighter and more attractive ones on the GSM network also contributed to this.[425]

**2. Globalstar**

Since early 2000 Globalstar became the second operational big Leo S PCS, starting providing services on 9 January 2000 Globalstar became a cellular fill-in for international or in-country users.[426]Due to the increased number of gateways, it has teamed up with a large number of local providers.[427]Globalstar has had serious launching problems, since it lost 12 satellites at launch. It was forced to delay its launch for the year 2000, although it was supposed to enter the market at almost the same time as Iridium.[428]

Globalstar had observed the moves of Iridium since the beginning, as it had yet to become operational. The fact that powerful investors in all three countries of

---

[424] See for a more analytical presentation, Celestial ruination, Public Network Europe, October 1999, Theresa Foley, Iridium and Globalstar hit delays, Communications Week International, 21 September 1998.

[425] On Iridium's history and critics see, Celestial Ruination, Public Network Europe, p.65, October 1999, Martyn Warwick, Mobile satellite: up in smoke, Communications international, 1 October 1999, Vanessa Clark, non immediate solution for troubled Iridium, in total telecom news of 26 April 1999, to be found in www.totaltele.com. See Iridium delays paging service until Mid-November, Satellite Week, vol., 20 no 44, 2 November 1998, Robert A. Nelson, Iridium, from concept to reality, Via Satellite, September 1998, p. 62, Carney, M. G., Strategy in the Telecommunica-tions market, the effects of liberalisation and competition, in Telecommunications Policy, September, 1986, p. 2245 ff.

[426] See Cellular globalisation dampens prospects for mobile satellite, Satellite international, 1999.

[427]Iridium had characterised Globalstar as a "localstar", which meant that it would not be able to provide global services. Satellite Week vol. 20, no 44.

[428] They will become operational after 2002.

the Triad backed Globalstar was already a factor for success. The bad course of Iridium made Globalstar a very careful market player. After Iridium's filing for Chapter 11, Globalstar became even more careful so as not to make crucial mistakes that would lead to failure. Globalstar based its market strategy on the philosophy of the following principle: to exploit fully the potential of the system's technical capabilities as well as those of the strategic partners, minimising the technological risks, and offering low-cost and high quality services. On the basis of this principal, Globalstar selected as partners very strong service providers and manufacturers in the EU, US and Japan, providers and manufacturers with long expertise in this field. For instance, selected service providers are Air Touch, Dacom, France Telecom, Vodafone and telecommunications equipment manufacturers Loral, Qualcomm, Hyundai.[429]To keep cost low, the systems architecture was based on a "bent pipe" structure, relaying signals received directly to the ground.

Furthermore, Globalstar aimed at developing strong relations with the local authorities and governments, which had provided affordable ground infrastructure. As regards the cost and quality of the prices, Globalstar was less expensive than Iridium. The service providers would be free to establish their own retail prices according to the local market conditions and regulatory requirements. According to Globalstar, further qualities of the system are its world-wide coverage, its superior call quality, no voice delay and multiple satellite coverage. Globalstar predicted approximately 336.000 subscribers by the end of the first year of service. Although Globalstar has a strong European participation (Vodaphone, Airtouch, France Telecom, Alcatel), it does not consider the European market as a primary one. [430]Furthermore, according to Globalstar, the international business traveller market, which Iridium aimed at, was in fact smaller than expected, and it was necessary therefore to adapt a strategy that aimed more aggressively at other market groups.[431]

Initially, Globalstar only offered services on a limited basis and global coverage was expected before April 2000. The market of Globalstar focused on voice services, mainly for the following categories: first, fixed subscriber terminals for areas lacking adequate terrestrial infrastructure, second, mobile subscriber ter-

---

[429] In particular for the development of the space segment Loral and Qualcomm are responsible, Aerospatiale Espace and Defence, Alenia Space, Daimler-Benz Aerospace, Hyundai. Launch service providers have been Mac Donned Douglas, NO Yuzhnoye, China Great Wall Industry Corporation. Qualcomm was responsible for the development of the gateways and the terminals.

[430] Chris Forrester, LEOs, GEOs, MEOs; Will they fly?, Via satellite October 1997, p. 26.

[431] See for forecasts on Globalstar sales www.satellitetoday.com/snapshot/index.htm.

minals as a cellular complement, using dual mode phone for global and local roaming and third, other services like messaging and paging.[432]

However, despite good perspectives, things did not turn out to have a better ending. Despite the lower prices, customers did not trust the mobile satellite network. The main reason for this being Iridium's failure. As a consequence, the customer's base was very limited and revenues were very low. These severe financial difficulties forced the system to file for Chapter 11.

### 3. ICO

ICO enjoyed the solid support of many European investors and INMARSAT shareholders, since ICO started as an INMARSAT affiliate. This was considered to be particularly important for ICO's success in Europe. Furthermore, ICO as a late starter, it expected to benefit from Iridium and Globalstar.[433]Moreover, ICO claimed to be the system with the lowest costs and the biggest cost-effectiveness because of the following reasons. First, the MEO orbit would eliminate the danger of signal blockage of the LEO systems, as well as the propagation delays of the higher GEO orbit. Second, as the orbit would be higher, it would mean that fewer satellites in the constellation would lower the cost of space segment, as there would also be no need for intersatellite links. Third, ICO partners would be telecom service companies that had an already firmly developed customer basis, something that would guarantee experience.

Like the other S PCS, ICO had identified a wide range of potential customers: cellular users that wanted to be served in areas where there was no or incompatible coverage; aeronautical, maritime and transport operators and populations in remote and rural areas. ICO would offer digital voice, data, fax and messaging services. It would also be a dual mode phone, compatible with Japanese and US standards. ICO focused on two major markets: First, international business travellers, requiring global roaming capacity; and second, local and regional business travellers for local roaming at a lower cost. All other segments of cellular complement and fixed wireless services would also be covered. The service will be fully complementary with terrestrial cellular and PCS.[434]

All these reasons would have guarantee success, but ICO, because of the fund raising difficulties, filed for Chapter 11 against bankruptcy.[435]With the financial

---

[432] Euroconsult, World Satellite Communications and broadcasting market survey, Prospects to 2006, Report, 1993.
[433] Ibid., B-17.
[434] Ibid., B-16.
[435] Before filling for Chapter 11, ICO had proven compliance with the 5 first milestone criteria.

assistance of Teledesic's investor, ICO managed to overcome the difficult situation and find investment, in order to restructure and transform into an internet-based service provision. ICO-Teledesic is preparing to enter the market in 2005 providing broadband satellite applications.

### *IV. Regional S PCS*

The factor of geographical coverage changes the strategies of the various Big LEOs, as it differentiates between global and regional systems. Many of the regional systems proposed have not only been LEO systems, but also GEO ones. The global LEOs, Iridium, ICO and Globalstar have been aiming at providing services all over the world (with the exemption of polar areas in Globalstar's case). ECCO, Ellipso, East and the other regional systems would focus mainly on the needs of regions, be they equatorial countries or continents like Africa, Middle East, or Latin America. Compared to the other LEO systems, they would be low cost. They targeted market groups such as businessmen, government agents and other individuals that would need either mobile coverage or fixed connection in rural areas to provide services like voice, data, fax, voicemail, position determination, call forwarding and call waiting.[436]Moreover, they aimed at domestic telephony and their prices would be lower than the ones of the global big LEOs.[437]For instance, ECCO decided not to cover the whole world, preferring instead to have satellites above the equator and provide coverage between the Tropic of Cancer and the Tropic of Capricorn, an area that includes most of Latin America, Africa and Southeast Asia.

This strategy of regional systems focused on countries that did not yet have extensive copper wire or fiber systems and that would therefore offer a far easier market. For instance, in ECCO's coverage area there were fewer than two phone lines for every 100 people in 2001. This strategy has been called "developing world strategy". However, pricing has been an issue. People in these areas could not afford that much and international travellers in the desserts and rural areas were not enough to constitute a critical mass market that could guarantee renevues for the systems. Globalstar and Iridium have planned systems of billions of dollars, whereas ECCO was looking only to operate on a limited basis, which if it worked properly could be easily expanded.[438]Regional coverage guaranteed quality, because the service concentrated on a specific area and it was cheaper and so ECCO was especially suitable for someone who used a satellite phone in the area of the equator.

---

[436] See Ovum, LEOs, MEOs and GEOs: The market opportunity for mobile satellite services, Report, 1998, p. 246.
[437] $1-$0.50 per minute.
[438] Spiegel, Let the big boys go first, Forbes Magasine, October 19, 1998.

*V. Little LEOs*

Little LEOs would provide communications to fixed and mobile points, transport, remote sensing, monitoring and positioning and would aim at providing services for personal, business, government, health, education, industrial and environment applications in many market segments.[439]

They would, moreover, be the cheap alternative to big LEOs, providing seamless services. They would not concentrate on voice applications, but on fleet and cargo management, data collection, security and property management, emergency, personal communications, and remote operating functions. Moreover, they would provide specific market applications, concentrating on data services, whereas Big LEOs concentrate mainly on voice applications and then data. This would mean that if a user wants data services, he was likely to choose Little Leo services.

Little LEOs have emerged as systems of advanced micro-satellite technology and low mission costs. They have been global systems that provided narrowband two way digital services for personal business and environmental and industrial services. They appeared to provide the low cost niche in the telecommunications service that benefited populations,[440]targeting a great variety of customers with a variety of applications that are specifically data based. The Little Leo Market was expected to be worth $5.5B by 2002. Although the Big Leo market overlapped with the Little Leo market, Orbcomm operators, (which has been the first in use) were very optimistic and ambitious about the future success of the system. According to the marketing strategy of Orbcomm, users were going to see it as a wireless e-mail suitable to all people that could not afford the prices asked by Iridium. The biggest profit of its market would come from the two-way messaging service.

The main markets that Orbcomm targeted were markets that included transportation, commercial marine, oil, utilities, personal messaging and government services. The company had made agreements with 15 regional providers to offer their services around the world. Those regional providers had, until the end of 1998, distribution agreements with more than 115 companies in the 22 countries

---

439 For agriculture, the system enables remote sensing and facilitates remote control of irrigation. For environmental applications the systems supports remote filed data collection from sensors, such as volcano tilt meters. In the transportation industry the system enables tracking trucks, shipping containers, and rail cars. Michael H. Ahan, Little Leo Systems and Services, as above, 1998.

440 Michael H. Ahan, Little Leo systems and Services: Accessibility with national sovereignty Control Brings Benefits to Asia, Global Communications, Asia 1998, Edition Hanson Cooke, 1998.

licensed to offer the service. Orbcomm hoped to obtain licenses for service provision to 80-90 countries by the end of 1999. It also expected to have about 30% of its business coming from the two-way messaging market.[441] However, like Globalstar, and despite being a Little Leo, their very limited customer base led to severe financial difficulties and they filed for Chapter 11.

### *VI. S PCS and market assessment*

### 1. Competing to fail? General Estimations

Systems were optimistic about their success for several reasons. With regard to technology they assumed that the untested technology would not be a problem. Systems would work without any difficulties and if there were any, these could be easily solved. However, the untested S PCS network technology has created considerable risks for the industries that have participated; launching failures and satellites that never reached their final orbit position have been common hazards in space operations. The more complicated the technology, the greater the expense for the consortium. Systems that used intersatellite links needed more investment and consequently had to raise prices for the services offered. Iridium, for instance, had as a start up price $7 per minute, a rather unattractive to customers price.

In terms of space in the market interested parties thought that there would be enough for all systems, as there would be enough spectrum available for all interested participants and there would be an increased number of customers (optimistic market forecasts predicted many millions of users). In addition, there was the issue of different marketing strategies, related to many important issues. These have in particular, involved the choice of gateway operators and service providers, identification of ways of getting nearer to the end user, and identification of group segments in each country. The more experienced these partners would be, the more success could be guaranteed since these would have been the entities that would have a direct interface with the national administrations and the end users. The systems seemed to have neglected this factor, as they focused a lot on advertising strategies. In the end the partners chosen were largely inexperienced.

In terms of competition in the market investors claimed that there would not be competition among the systems, since each system would target different market segments. Each system gave a different name for the same user category, i.e., the ones outside GSM coverage, international travellers, or users in rural areas

---

[441] See Wireless Europe, 19 October 1998. E-Sat and VITA have already received a license from FCC, at the same time as Orbcomm. However it seems that they are facing technical difficulties. More companies have in addition have filed applications. These were CTA, Final Analysis, GEAmericom and LEO One.

etc.[442]With regards to services, Big Leos promised to offer both voice and data services, whereas Little Leo would specialise mainly in data services as well as specialised applications like maritime, trucking, rural telephony and remote monitoring, which would play a much more significant role and giga leos would offer broadband applications.[443]However, all systems targeted markets with the same needs, i.e., users in areas where mobile terrestrial coverage was very weak or even non-existent, something which could lead to market overlap.

In terms of rival (GSM) technology all systems hoped to attract mobile cellular customers. In the non-rural areas the GSM related standards were dominant and in the rural areas S PCS operators estimated that they would be able to reach a critical mass. However, none of the systems took into consideration the huge growth of the GSM market, which definitively limited their opportunities and profits.[444]S PCS operators, having seen the success of the cellular markets, tried to create a ground infrastructure that would have similarities with the terrestrial cellular one and the services that the S PCS would offer would be more or less the same as the ones cellulars offer, with the difference being the global coverage. The existing and emerging S PCS were promising compatibility with the existing cellular technologies.[445]Iridium, for instance, wanted to offer dual mode phones that would allow the user to switch to the cheapest GSM tariff without dropping the call and ICO promised pocket-sized units that would be of dual-mode, capable of working with satellite, cellular and PCS systems based on GSM, CDMA, D-AMPS and PDC.[446]

---

[442] PNE, October 1998, p. 36.

[443] PNE, October 1998, p. 35.

[444] There have been various important players that have contributed to this success through innovative technological applications and clever marketing strategies. First, powerful industrialist manufacturers like Ericsson, Nokia, Siemens, Alcatel, Panasonic, Motorola, Matsushita, Mitsubishi, NEC (the three biggest Japanese manufacturers). Second, network operators like DT, Airtouch, Mannesmann, STET, BT, Cable and Wireless, FT, Belgacom, Mobistar, PTT Telecom, Tele Denmark, AT&T, Telia, NTT DoCoMo, Vodaphone Group, Bell South, Sprint, Telefonica, Orange, One-2-One. Third, service providers, such as DT, T-Mobil, Omnitel, Panafon, e.t.c. European industrialists have been following various strategies, for instance planing according to the regulatory constraints and the demand, focusing not only on one technology and expanding: Latin America, Asia and of course towards Eastern Europe, where they see a market of 16 countries. There has been in fact a penetration of West European and US operators and providers as shareholders in the main Eastern European Operators, as well as suppliers. For instance: US West/Bell Atlantic (USA) 49% in Euro Tel Praha, PTT Nederland, Tele Denmark, Telecom Finland, Telenor, Telia 74% in Pannon (Hungary), France Telecom 34% in Centertel (Poland), US West 22,5%, T-Mobil 22,5% of PTC Poland, Tele Denmark (Denmark) 19,3%, AirTouch 19,3% of Polkomtel ( Poland) Ericsson 36%, Alcatel Telecom 12%, Nortel 2%, Motorola 12%, Siemens 12%, Nokia 26%.

[445] For instance., MSS/GSM, AMPS/CDMA/MSS, PDC/MSS.

[446] All are transmission modes, based on different protocols.

The S PCS that went operational claimed not to compete with mobile cellular operators. Cellular operators have been amongst the largest investors in the S PCS, which meant that for them the more the customers the bigger the profits. The market would in theory be open and competition was expected to be intense, as S PCS would battle as hard for market share as their terrestrial partners.[447]The success and failure of the S PCS systems has been determined by a number of factors. First, the price of the terminals and the monthly fees: if S PCS wanted to compete with or complement the terrestrial networks, the prices had to be at some point comparable. Second, the quality of services: GSM users have been very satisfied with the services and it has been be very difficult to persuade them to change, unless the relevant services offered a higher quality. Third, the appearance of the terminals: S PCS terminals are bigger and heavier compared to the existing cellular ones and so the customer preferred cellular terminals. Fourth, the usage in building: cellular operators have increasingly been encouraging the substitution of cell traffic from fixed to mobile networks.[448]

**S PCS planned market development scenario**

| | |
|---|---|
| Applications | Initially voice would be dominant. However, due time, data would become increasingly important, internet being the main application |
| Service availability | Almost global coverage |
| Key technology developments | Satellite would be the complement to the terrestrial wireless applications of 3G. |
| Pricing | Prices would be dependent on the service provider and applications. They would follow a gradually descendant course, ending up at prices similar to the ones of GSM. |

What has happened to the market? During the last years dozens of S PCS project have made their appearance, many billions of dollars have been spent on technology designs, subsequent bankruptcies have taken place, investors have been quitting projects and consumers have been reluctant to try the new services. The spectacular failures, as initiated by Iridium, have put a sort of "death label" on the industry raising numerous doubts about the purpose of the market's existence.

[447] Mobile via Satellite: bent pipe dreams?, December 1997, p. 48.
[448] See European Commission, Satellite Action Plan, SEC (1999) 250, 16.2.99.

Market mechanisms have triumphed revealing one of the biggest business failures in the telecommunications market. Systems that became operational were in constant need of capital and in the end were sold a few millions to asset auctions. The reselling and the auctioning aimed at the readaptation of managerial plans, the improvement of the system's services as well as the lowering of the prices. However, none of these changes was enough to attract more customers.

The ones that were supposed to enter the market later on either have withdrawn or have become very careful observers, postponing the starting date and trying to integrate new advanced applications in order to make their services more attractive to potential customers. Despite the big disappointment there is still interest in satellite systems and in particular broadband applications. Industrialists acknowledge that the need for broadband will increase, as consumers will be looking for affordable broadband services. Satellite projects that were planned as broadband, like for instance Teledesic, and not narrowband, have more technical possibilities for advanced applications. However, difficulties still exist.

**2. S PCS, satellite industry and challenges**

Satellite industry has been in the process of restructuring as during recent years it has witnessed the break up, buying and selling as well as privatizing of companies.[449]The market will grow, but forecasts see satellite technology being mainly used for broadcasting and broadband applications. Neither the mobile high-speed internet projects nor, the geostationary broadband satellite projects are evolving the way in which they were expected. Many investors and providers have withdrawn from planned projects, others are far from making profit. Investment failures for narrowband LEOs, such as Iridium, ICO and Globalstar have been disappointing examples for the industry, as they have made potential newcomers very sceptical as far as the technological complexity of the project, the financial requirements, the market demand and profit generated are concerned. Fixed satellite operators are in a better position than the mobile ones, who during during the past ten years have been struggling for financing.[450]Mega projects, such as Teledesic and Globalstar aimed at establishing satellite multibillion business through internet and multimedia applications, however they remained stagnant as all projects had to be reconsidered due to financial requirements, overestimated demand and competition from rival technologies.[451] Two-way satellite operators offering internet access are suffering. However, the

[449] Foley, Satellite Sruge, Communciations Week International, 24 September 2001

[450] David Molony, Consolidation or collision, Communciations Week international, 24 September 2001, Foley, Business Strategies: Satellite companies: gasping for oxygen?, Communications Week International, 16 April 2001.

[451] Taaffe, Two-way satellite operators aim at SMEs, Communications Week International, 6 May 2002.

industry is optimistic for the internet satellite market and in particular, regional markets. Despite the scepticism, broadband applications promise a bright future and, in general, the business of satellite operators continues to grow as the demand for the internet through transponder increased.[452]

**3. Broadband Multimedia Satellites**

There have been, in particular, various S PCS projects aiming to develop infrastructure in order to provide broadband multimedia applications, which will be developed to support a market based on a common network platform, an open network based on common standards and protocols, following the 3G evolution.[453]However, the broadband satellite market has been following a slow pace of evolution mainly for the following reasons. First, because of the technical difficulties that operators had to face up; second because of the high costs for smaller service providers to buy capacity for the end user services and third, the satellite sector could not oversee the investment failures of Iridium, ICO and Globalstar.

---

[452] See Broadband Satellite: analysis of Global Market Opportunities and Innovation Challenges, January 2002, Pionner Consulting.

[453] Broadband are services which simultaneously enable internactive operation and a transmission speed of at least 2 Mbit/s, see Omnitele, Study on the technical and regulatory requirements, p. 12. Multimedia is commonly associated with the combined presentation of data, sound and vision in software applications stored in CD-ROMs. A more appropriate definition is "services arising from the convergence of computing, communications and established media" in Susan Ablett and Simon Norris, Multimedia in Telecoms (Cambridge 1994). Also Analysis, UMTS Market Forecasts, 1997. Satellites are particularly promoted for internet net applcations. The broadband multimedia satellite market is based on the belief that satellites, as a solution to the terrestrial alternative for itnernet services are better for the follwing reasons. First, they provide high capacity compared to many other transmission technologies. Second, they allow bandwidth to be allocated at a variable rate. Third, it is well suited for emerging broadcast applications. Cynthia Boeke, Surfing among the stards, Via Satellite, November 1997, Also Peter J. Brown. Global Solution now, IP over satellites, Via Satellites, october 1998, p. 17. Also Struzak, R., internet in the Sky: the tests have started. Global Communmciations, Asia 1998, Hanson Cooke, Omnitele, Study on technical and regulatory requirements for open access to broadband telecommucation networks and services for customers, service providers and content providers, final report, June 1998, p. 5. Also DTT Consulting, Internet Via Satellite, 1998, p.61 Broadband are services that simultaneously enable interactive operation and a transmission speed of at least 2 Mbit/s, see Omnitele, Study on the technical and regulatory requirements, p.12. Multimedia is commonly associated with the combined presentation of data, sound and vision in software applications stored on CD-ROMs. A more appropriate definition is "services arising from the convergence of computing, communications and established media", in Susan Ablett and Simon Norris, Multimedia in Telecoms (Cambridge, 1994).Also see Analysis, UMTS Market Forecasts, 1997.

There have been fears for the realisation of the potential of the market, as the systems have had to compete with other already established and successful technolgies, such as the digital subscriber lines and the cable modems. What providers had to do was to carefully calculate the capital needed for R&D, the expected demand and the revenue that will be generated by this demand. Investors have been trying to see which services would generate the desired revenues as all broadband projects have been uncertain as to how large the market will be.[454]

Broadband multimedia satellites aim to support a market based on a common network platform, an open network based on common standards and protocols, maintaining the same structure of distributed network architecture as the S PCS. Moreover, they aim to target customers who need broadband connections for internet access, computer networking, aggregation and trucking of voice lines and telecommuting, services which become expensive outside the highly populated areas, expensive or simply unavailable at any price.[455]Many ambitious projects have been launched -Teledesic, Cyberstar, Astrolink, Horizons being some of them. However, once more, the risk of untested technology, coupled with the issue of financing has forced many to withdrawal and others to postpone the year of launching.

---

[454] Foley satelite surge, Communications Week International, 24 September 2001.

[455] ICO, ICO TENS: The potential for pan-European multimedia mobility, 1999.

## Conclusions - Summarising Remarks

Through decades of technological evolution, satellites have shown that they are able to provide service solutions, either additional to the existing terrestrial services or without the intervention of terrestrial services. However, the process of designing, launching and operating a satellite is complicated. In order to launch a system and enter the market to provide services, technology, primarily, needs to be developed. The satellite constellation, as well as the whole infrastructure, (gateways, ground control stations and mobile terminals) has to be designed. Such a project is initiated by a company or a group of companies which, parallel to the manufacturing of the satellites, tries to find and attract investors-partners. It takes years to realise such a project, since numerous obstacles must be overcome from a technical, financial, administrative and regulatory point of view. The more investors a satellite project attracts, the greater the possibilities it has to survive. There have been projects, which due to technical and financial difficulties, have been abandoned (e.g. Odyssey). Others, like Iridium and Globalstar, initially had to face delays, became operational, but with a small customer case, were sold in auctions and are just surviving, still facing financial difficulties.

Technology facilitates the provision of services. However, it is the law that allows market players to enter the market, trade products, and provide services, and that facilitates their interaction. Law facilitates the entry and guarantees the integrity of the players as well as the customers' right to qualitative services. Law makes trade easier, through the abolishment of trade barriers, the sinking of licensing fees, and custom duties. It may guarantee the success of procedures for setting prices and performing spectrum assignment procedures, authorisation procedures or indicate how S PCS systems should be examined before being licensed in a European state. The most difficult market entry conditions are those of the assignment of frequencies and the granting of operational license by the states, where the system wants to provide services. The assignment of frequencies is made by the state to which the system first applies and is represented at the ITU level. Neither internationally, nor in Europe are there laws that oblige states to use a certain frequency assignment method. This is because of the specificity of the matter and its economic importance for the budget of the state. For further licenses in other states, the system has to respect the provisions of each national legal order. At a European regional level there has been an attempt to harmonise the national legal orders on licensing monitored by the European Commission and CEPT based on documents such as the Licensing Directive, ERC/DEC/(97)03, ERC/DEC/(97)05 and ECTRA/DEC/(97)02.

In this framework, economic actors built their strategies. There has been an increased number of interested players. The level of success of the S PCS system has been determined by a number of factors.[456]These have been the following. First, the level of demand: the system had to attract enough users in order to have sufficient revenues. Second, the service cost. Business users that would be able to afford it would not be that many to constitute a mass market. Third, the risk factor: where issues of satellite infrastructure, software and insurance were at stake. Fourth, service quality: Because of the already existing successful GSM, customers would expect the same high level of service. Fifth, political or projectionist obstacles that would prevent mobile satellite operators from accessing markets.

Technology risks and the difficulty in finding strategic investors that would be willing to invest billions and wait for years to see some profit have been two major factors of business failure. Moreover, the whole procedure of making an S PCS operational has been time and money consuming. PCS systems had to go through many difficult phases: successful designing and launching of the satellites, development of proper marketing and distribution strategies, getting operational licenses from many different countries.[457]In addition, given the high growth of the mobile terrestrial telephony, it was made clear to the S PCS investors and operators that the success of the systems would by large depend on the relationship between the satellite and the terrestrial mobile systems. This was the reason why the job of finding mobile partners (either as investors, manufacturers, or providers) or roaming partners constituted one of the principal activities of the Systems. Moreover, this was why all systems based and promoted their strategies on roaming facilities that S PCS systems would offer and all systems wanted to exploit the fragmentation caused by the incompatibility of terrestrial standards, and as well as their geographical limitation. However they knew that satellite transmissions could not compete with cellular systems in urban areas, and that it would be very difficult to attract customers and this was why they were arguing that satellite and cellular systems were complementary.

The aim of conquering the S PCS market has been an important challenge. However, these systems have been experiencing a major problem. This problem has not been that of regulatory entry barriers or marketing strategies, but mainly the raising the financing needed to proceed. In addition, careless marketing moves, like that of Iridium (which spent more than $140m on the advertising campaign), lead to a lack of funds and the inability to pay back debts. The so-called "Iridium flu“ has harmed satellite projects, by causing delays or even can-

---

[456] Public Network Europe, Yearbook, 1997.

[457] Chris Forrester, LEOs, GEOs and MEOs, will they fly? Via Satellite, October 1997, p.20.

cellations, as companies started withdrawing from the systems because of lack of financing and a reduced degree of confidence.[458] In fact, the fears of many analysts ended up true, wondering if rivals would be in danger of burning up after the collapse of Iridium.

The case of Iridium has shown that systems may enter the market with the best perspectives but end up filing for bankruptcy protection. Up to now, the majority of the systems have had fund raising problems. The complicated and untested technology became a dangerous risk factor for the investors. Lost satellites in orbit, gateways and terminal technical problems make the investment of bigger capital necessary. This discouraged investors, who had to invest for many years before they could see any profits. Proof of this has been the very low percentages of the customers, much lower than expected, for all projects, not only the global but also the regional ones.[459] The same is happening in the satellite broadband multimedia systems. Here the technology risk factor is even greater, since the technology is even more complicated. Up to what degree the new satellite systems will be threatened, is something that only the market will show.

---

[458] Aviation Week and Space Technology, as above, p. 152, also Geneviever Wilkinson, Reuters, Satellite companies face murky fate after iridium's demise, in totaltelecom news of 27.3.00.

[459] Thuraya, for instance, had 7000 phones sold in the first six weeks of operation and commercial service provision. The system expected however to have 400000 subscribers in the first year of operation. Theresa Foley, Satellite surge, Communications Week International, 24 September 2001.

# BIBLIOGRAPHY

## Monographs

Abdelaziz M. (ed.), Les alliances strategiques dans les technologies de l'information, Paris, 1996

Ablett S. and Graham J., Strategies for Global Networks, Analysis publications, Cambridge UK, 1993

Ablett S. and Simon N., Multimedia in Telecoms, Cambridge, 1994

Abolafia M. Y., Market as a cultural approach, in Hodgson M. G., Screpanti E., Rethinking Economics, Markets, Technology and Economic Evolution, Aldershot: Elgar, 1991

Abolaffia, M. Y. and Biggart, N. W., 1991, Competition and Markets: an institutional perspective, in A. Etzioni and P. R. Laurence (ed.) Socio-economics: Towards a new synthesis, Armonk, NY: ME., Shapre

Adamantidou E., La désetatisation des services d'utilité publique, Diss. EUI, Florence, 1996

Ahan M. H., Little Leo systems and Services: Accessibility with national sovereignty Control Brings Benefits to Asia, in Global Communications, Asia 1998, Edition Hanson Cooke, 1998

Albrow M., The Global Age, Cambridge: Polity Press, 1996

Albuquerque J., Hayden T.L., WRC-97 Completed Regulatory Framework Enabling Non-GSO FSS Systems in 2x500 Mhz of Centimetric Waves (Ka-Band), in Global Communications, Asia 1998, Hanson Cooke

Amsden Alice, H., 1992, A theory of government in late industrialization. In L. Patterman and D. Reuschemeyer (eds.), State and Market in Development: Synergy of Rivalry, Boulder CO: Lynne Rienner

Analysis, Comparative Overview of Current Regulatory Environment Telecommunications and Broadcasting Sectors, Report, January 1998

Analysis, The impact of non-European competitors on a deregulated European satellite telecommunications market, Report for the European Commission, January 1993

Analysis, Mobile Market Review, Report, 1997

Analysis, UMTS Market Forecasts, Report, 1997

Analysis, Adapting the EU Regulatory Framework to the Developing Multimedia Environment, Report, January 1998

Ananasso F., Vatalaro F., (ed.) Mobile and Personal Satellite Communications, Springer Verlag, 1995

Antonelli C, The economics of information networks, New York, Elsevier Science, Amsterdam, 1991

Antonelli C., Technological change and multinational growth, international telecommunications services, in Noam E. and Wolfson H., (ed.), Globalism and Localism in Telecommunications, 1997

Archibugi D. and Michi J., Technology, Globalisation and Economic Performance, Cambridge, U P, 1997

Arnbak J. N., Managing the radio spectrum in the new environments, in W. H. Mellody (ed.), Telecom reform, policies, principles and regulatory practices, 1997

Asuberl, J.H. and Sladovich HT., (ed.), Technology and Environment, National Academy Press, Waschington D.C., USA, 1989

Axel G., Voigt R., Rechtspolitologie, Eine Einfuehrung, Westadt Verlag, 1985

Axtmann, R., (ed.), Globalization and Europe: Theoretical and empirical investigation, London, Pinter 1998

Baldwin R., Cave M., Understanding regulation, theory, strategy and practice, Oxford University Press, 1999

Ball, T., Transforming Political Discourse, Political theory and Critical conceptual History, Oxford, 1988

Baldassari M., International economic interdependence, patterns of trade balances and economic policy coordination, New York: St Martin's Press, 1992

Balwin R., Towards an integrated Europe, London, Center for Economic and Policy Research, 1994

Bate S., White S., Johnson T., Satellite Communications in Europe, 2nd Ed., London, FT Law & Tax, 1996

Bar F., Borrus M., Coriat B., Information Networks and Competitive advantage: Issues for Government Policy and Corporate Strategy Development, Brussels: Commission of European Communities, DGIII. BRIE-OECD Research Program, 1991

Bar F. and Borrus M., The Future of Networking, Berkeley, CA: University of California, BRIE Working Paper, 1993

Barnett R. J., and Cavanagh J., 1994, Global dreams, imperial corporations and the new world order, New York, Simon and Schuster

Bartlett C.A., Ghoshall, S., 1990, Managing innovation in transnational corporations, London, Sage Publications

Bartlett C. A., Y. Doz, Hedlund G., (ed.) Managing the global firms, London, Routledge, 1990

Barton S.K., Introduction to CDMA, CNTE publication, August 1993

Bauer J., Steinfield Ch., Telecommunications initiatives of the European Communities, in Steinfield Ch., Bauer J. and Caby L., Telecommunications in transition: policies, services and technologies in the European Communmications, Thousand Oaks: Sage 1994

Beck U., Was ist Globalisierung, Suhrkamp, 1997

Beck U., Politik der Globalisierung, Suhrkamp, 1998

Becker L. J. E., (ed.), Business Netwerke, Stuttgart, 1996

Bell D., The Coming of the post-industrial society, a venture in social forecasting, 1973, Basic Books, Inc. Publishers, New York

Bellemans L., Tsoukalis L., Europe and global economic interdependence, Bruges European Interuniversity Press, 1993

Bender R., Launching and Operating Satellites, Legal Issues: The Hague, Nijhoff, 1998

Beninger J. R., The control revolution: technological and economic origins of the information society, Cambridge Harvard University Press, 1987

Bhagwati J. N., Irwin Douglas A. (eds), Political economy and international economics, Cambridge, 1991a.

Bieder R., Dehousse R., Pinder J., Weiler J.H.H., (ed.) 1992: One European Market? A critical analysis of the Commission's internal market strategy, Baden-Baden, Nomos, 1988

Biersteker, T.J., Globalization as a mode of operation: conceptual changes within firms, states and international institutions, Institute for international studies, Brown University, providence, RI June, 1994

Black J., Rules and Regulations, Oxford, Clarendon, 1997

Blandin A. and Abdelaziz M., Les alliances stratégiques des operateurs de télécommunications: nécessité économique et contrainte juridique, in Abdelaziz Mouline, Les alliances stratégiques dans les technologies de l'information, 1996

Bliss Ch., Economic theory and policy for trading blocks, Manchester, Manchester University Press, 1994

Bohlin E. and Stanford L., (ed.), Telecommunications transformation, technology, strategy and policy, Ios press, Amsterdam, Netherlands, 1997A

Bourély M., Mobile satellite telecommunications, in Lafferanderie Gabriel (ed.) Outlook on Space Law over the next 30 years, The Hague, 1997, 385-398

Bower, J.L, 1987, When markets quake, Boston, MY, Harvard Business School Press.

Bowett, D.W., The Law of International Institutions, 4th Ed., London: Stevens, 1982

Boyer R., State and the Market, in Boyer/Drache, States against Markets, The Limits of Globalisation, London, Routledge, 1996

Boyer R., Drache D., States against Markets, The limits of globalisation, London: Routledge, 1996

Bradley, St., Hausman J. and Nolan R. (eds), Globalization, Technology and Competition, Boston, Mass., Harvard Business School Press, 1993

Branscomb L. M., National Corporate Technology strategies in an interdependent world economy, in B. R. Guile and H. Brooks, (ed.), Technology and global industry, Companies and nations in the world economy, Waschington D.C, National Academy of Engineering, 1987

Broadband Satellite: analysis of Global Market Opportunities and Innovation Challenges, January 2002, Pionner Consulting.

Brock G. W., The telecommunications industry, the dynamics of market structure, Harvard University Press, 1981.

Brown S., International relations in a changing world system, Boulder, CO: Westview Press, 1992

Bruce R. R., Cunard P. J. and Director D. M., From telecommunications to electronic services, a global spectrum of definitions, boundary lines and structures, 1986, Butterworth

Caby L., Steinfeld Ch., Trends in the Liberalisation of European Telecommunications, Community Harmonisation and National Divergence, in Steinfeld et al., Telecommunications in transition, Sage Publications, 1994

Calliess Ch., Ruffert M. (ed.), Kommentar des Vertrages ueber die Europaeische Union und des Vertrages zuer Gruendung der Europaeischen Gemeinschaft, Darmstadt 1999

Camilleri J. A., Falk J., The politics of a shrinking and fragmenting world, Edmond Elgar, 1992

Carnoy, M., Manuel C., Stephen S. C. and Cardoso F. H.: The new global economy in the information age. Reflections on our changing world, University Park, PA: Pennsylvania State University Press, 1993

Carpentier M., Farnoux S., Toporkoff, G., Telecommmunications in transition, John Wiley Chichester UK, 1992

Castells M. and Hall P., Technopoles of the world: The makings of 21rst Centrury Industrial Complexes, London Routledge, 1994

Castells M., The Rise of the Network Society, Cambridge MA: Blackwell Publishers, 1996

Cawson A., Morgan K., Webber D., Holmes P. and Stevens A., Hostile Brothers: Competition and Closure in the European electronics industry, Oxford, Clorandon Press, 1990

Center for Economic and Policy Research, Europe's network industries: conflicting priorities: Telecommunications. Monitoring European Deregulation series, CEPR, 1998

Chioni G., Regulating Frequencies: Actors, policy-making and legislative framework in the filed of mobile and satellite communications, Aristoteles Public Law Series, vol.V, Esperia Publications, 1999

Chioni G., Globalisation of Personal Communications:The Satellite Personal Communications Case-Regulating a Global Interplay of Actors for a Market or a Failure?, PhD, European University Institute, Florence, 2000

Chomsky N., Dietrich H., Globalisierung im Cyberspace, Horlemann, 1996

Clark I., The hierarchy of state. Reform and resistance in the international order, Cambridge University Press, 1989

Coase R., The Firm, the Market and the Law, University of Chicago Press, 1988

Cocca, A. A., Towards an adequate legal regulation of the Geostationary orbit, in Proceedings of the 20th Colloquium on the law of outer space, 1978

Coclough, Ch. and Manor J., (eds), States or Markets? Neoliberalism and the Development Policy Debate, Oxford, Clarendon Press,1991

Common John, A sociological view of sovereignty, New York, Kelley, 1967

Commons John, Institutional economics, Band 1 and 2, University of Winsconsin Press, 1959

Communications Systems limited and Hogan & Hartson, Prospects for personal Satellite communications using service links in the Ka-Band, Report for the European Commission, August 1997

Connors J., The Race to the intelligent State: Towards the Global Information Economy of 2005, 1993

Connell St., Broadband Services in Europe, in Steinfeld et al, Telecomunications in transition, Sage Publications, 1994

Coopers & Lybrand, Study on analysis of new methods of frequency allocation in the Member States and comparative analysis of recent developments in this fields, July 1993

Corbert H. and Jackson R.,(eds), In search of a new economic order, London, Croom Helm, 1974

Coudert Brothers, Competition Aspects of interconnection agreements in the telecommunications secotr, Luxembourg, Report for the European Commission, 1995

Cowhey P. F., Aranson Jonathan D., Managing the world economy, The consequences of corporate alliances, New York, Council of Foreign Relations Press, 1993

Cox, R. W., A perspective on Globalization, in J. H. Mittelman (eds), Globalisation, Critical Reflections, edited by, 1996

Cutler Cl., Haufler V., Porter T., Private Authority and International Affairs, in Culter Cl. A., Haufler V., Porter T. (eds), Private Authority and International Affairs, State University of New York Press, 1999

Dahinden Erwin, Die rechtliche Aspekte des Satellitenrundfunks, Universitätsverlag Freibourg/Schweiz, 1990

Dalibor E., Handelspolitische Multilateralismus zwischen Globalisierung und Regionalisierung, , Berlin, Freie Universität, Diss, 1997

Daniels P.W., 1993, Service Industries in the World Economy, Oxford: Blackwell

De Sola Pool I., Technologies without boundaries, Harvard University Press, Harvard, London, England 1990

Deutch K. W., Sonntag Ph., From the industrial society to the information society: crises of transmission in society, politics and culture, Berlin, Wissenschaftszentrum, 1981

Deutsch K., The Nerves of Government, New York, 1963

Dezaley Y. and Garth G. Br., Dealing in virtue, International arbitration and the construction of a transnational legal order, The university of Chicago press, 1996.

Dicken P., Global shift: industrial change in a turbulent world, Paul Chapman Publishing Ltd, 1988

Dickson D., Εναλλακτική τεχνολογία, (Alternative Technology), Kalvos, 1985

Dietz A., Satellite Communications and Copyright Law, in Telecommunications Satellites and International Law, Colloquium, 1989

Dodgson. M., (eds.) Technology, Strategy and the Firm: Management and Public Policy, Harlow, Essex: Longman, 1989

Dracke D., From Keynes to K-Mart, in Drache, Boyer, States against Markets, The limits of liberalisation, 1996

Drucker P. F., Managing in a time of great challenge, New York, 1995.

DTT Consulting, Internet Via Satellite, Report, 1998

Dunning J., The globalisation of Business, London Routledge, 1992

Dunning J. H., Globalisation: The challenge or national economic regimes, The Geary Lecture, 1993a

Dunning J. H., Multinational enterprises and teh global economy. Addison-Wesley, 1993b

Durbin P. (ed.), Broad and narrow interpretations of philosophy of technology, Kluwer Academic Publishers, 1990

Dyson K., Humphreys H. (Ed.), The political economy of communications international and European Dimensions, Londres, Routledge, 1990

Eder K., Die Entstehung staatlich organisierter Gesellschaften. Ein Beitrag zu einer Theorie sozialer Evolution, Frankfurt A.M., Suhrkamp,1976

Eggertsson Th., Economic Behaviour and Institutions, Cambridge Surveys of economic literature, New York, Cambridge UP, 1990

Elbert B. R., Introduction to Satellite Communication, Artech House, 1987

Ellger R., 1992, Telecommunications in Europe: Law and policy of the European Community Law in a key industrial Sector, in Adams W.J. (ed.) Singular Europe: Economy and Policy of the European Community after 1992, Ann Arbor, The University of Michigan Press, 1993

Engelhard Helmut, Satelliten direktfernsehens im Spannungsfeld von Völkerverständigung und Propaganda, Frankfurt a.M., Peter Lang, 1992

ERO, Input into the preparatorial meeting for the ITU "First World Telecommunications Policy Forum ", 2 February 1996, London.

ERO, Spectrum below 2,5 GHZ, Report, Copenhagen, October 1997.

ERO, Mobile Broadband Systems (MBS), Report, Copenhagen, 1997

ERO, Mobile Satellite Services, Applications, Final Report, Copenhagen, 1995

ERO, DSI Phase III, Market scenarios and views on the development of radio services
within the range 862-3400 MHz, A consulatitive document to support the Detailed Spectrum Investigation Phase III, Copenhagen, October 1998

ETO, Final Report on Numbering Related to Personal Communication Service (PCS) in Europe, November 1997

ETO, Annex 15 to the Final Report on the licensing of Satellite Networks and Services, October 1997

ETO, Final report on licensing, February 1998

Euroconsult, World Satellite Communications & Broadcasting Markets Survey, Prospects to 2006, 1996

Euroconsult, World Satellite Communications and broadcasting markets survey: prospects to 2007, Paris: Euroconsult, 1998

Euroconsult, Economic impact of competing space segment, Final Report and Summary, March 1993

Europe, Mobile Satellite Services, March 1998.

Faulhaber G. R., Telecommunications in turmoil: Technology and public policy. Cambridge, MA: Ballinger, 1987

Fawcett J. E. S. and Audrey Parry, Law and international resource conflicts, Clarenton Press, Oxford, 1981

Fawcett J., E., S., Outer Space: New Challenges to Law and Policy, Oxford Clarendon, 1984

Featherstone M., Lash S., Robertson R., (eds.), 1995, Global Modernities, London and New York, Routledge, 1995

Featherstone M., (ed.), Global culture: Nationalism, globalisation and modernity, London Sage, 1990

Feenberg A., Critical theory of technology, Oxford, UK, Oxford University Press, 1991

Ferrazzani M., Soft Law in Space Activities, in Gabriel Lafferanderie, Outlook on Space Law in the Next 30 Years, The Hague, 1997

Fesenmair, J., Öffentliche Dienstleistungsmonopole im europäischen Recht, Berlin Verlag, 1995

Fieten R., Globalisierung der Märkte, Schaeffer Poeschel, 1997

Forester T., High Technology Society: the story of the information technology revolution, Cambridge MIT Press, 1987

Foster C., Privatisation, Public Ownership and the regulation of natural monopoly, Oxfor 1992

Fouries Fr. N., The nature of the market: a structural analysis, in Hodgson M. Geoffrey, Screpanti Ernesto, Rethinking Economics, Markets, Technology and economic evolution, 1991

Freeman C., and Hagedoorn J., Globalisation of technology, 1992, Brussels, EC-Fast

Friedmann W., the changing structure of internaitonal law , New York, Stevens 1964

Friedmann, W., Legal Theory, Columbia University Press, New York, 5th Edition, 1967

Friedmann W, Law in a changing osciety, London Stevens and Sons, 2nd Ed, 1972

Fuchs G., "The telecommunications" highway for Europe after 92 or paving a dead-end street?: the politics of pan-European telecommunications networks and development, Working Paper 301, Max-Planc Institute für Gesellschaftsordnung, Köln, No 6, 1993

Fudge R., Gardiner J.G., The future of third generation mobile systems, Artech House, 1995

Fundesco, Telecomunicaciones 1991, Tendencias, Comunicaciones por Satélite, Informed Anuales de Fundesco, 1991

Gabriel J. M., Worldsviews and theories of international relations, London-Macmillan, 1994

Gardiner J.G., West D., Personal Communication systems and technologies, Artech House, 1998

Gebhardt A., Heinritz J., Wissner M., (ed.) Europa in Globalisierungsprozess von Wirschaft und Gesellschaft, Franz Steiner Verlag, Stuttgard, 1998

Gehring G., Geostationary Orbit, Technology and Law, in proceedings of the 19th

Colloquium on the Law of Outer Space, 173, 1978

Giddens A., The consequencies of modernity, Polity Press, 1990

Graham A, (ed.) Government and Economies in postwar world, New York: Rouledge, 1990

Grewlich K., Transnational enterprises in a new international system, Alphen aan den Rijn: sijthoff & Nordhoff, 1980

Grewlich K. W., Europa im globalen Technologiewettlauf - Der Weltmarkt wird zum Binnenmarkt, Guetersloh: Bertelsmann 1992

Grewlich Klaus W, Konflikt und Ordnung in der globalen Kommunikation, Wettstreit der Staaten und der Unternehmen, Nomos, Baden-Baden, 1997

Gruebler A., Technology and Global Change, Cambridge University Press, 1998

Guile B. and Brooks, H. (eds.), Technology and Global industry: Companies and Nations in the world Economy, Waschington D.C: National Academy of Engineering, 1987

Henderson J., The globalisation of high technology production, Routledge London, New York, 1990

Hills J., Deregulating telecoms: competition and control in the united States, Japan and Britain, London: Pinter 1986

Hirst P. and Thomson Gr., Globalisation in question: the international economy and the possiblities of governance, Cambridge: Polity Press, 1996

Hodges M., Multinational corporations and national government: a case study of the UK's experience 1964-1970, Farnborough: Saxon House, 1974

Hodgson G., M., Screpanti E., Rethinking Economics, markets, technology and economic evolution, Aldershot: Elgar, 1991

Hodgson G. M, 1988, Economics and Institutions, A manifesto for a modern institutional econmics, Philadelphia P.A., Univesristy of Pensylvania Press.

Horsman M. and Marschall, Andrew, 1994, After the national state, London, Harper Collins

Horváth Ch. and Merzrath J.,(ed.) Globalisierung, Der Schritt in ein neus Zeitalter, Springer, 1997

Howells J., and Wood M., 1993, The Globalisation of technology, London Belhaven Press

Hulsink W., Privatisation and liberalisation in the European Telecommunications: comparing Britain, the Netherlands and France, London New York, Routledge, 1999

Hulsink W., Do nations matter in a globalised industry? The restructuring of the telecommunications governance regimes in France, the Netherlands and the UK (1980-1994), Delft: Edurnon, 1996

Humbert M., (Ed), 1993, The impact of globalisation on Europe's firms and industries, London, Pinter Publishers

ICO, ICO TENS : The potential for pan-European multimedia mobility, ICO Global Communications, Report, 1999

IDADE, Study on the international market for digital communications: export opportunities for EU institutions, Report, 1997

International Telecommunications Union, ITU: A hundred years of cooperation, Geneva, 1965

International Telecommunications Union, WARC-92, Final Acts, in www.itu.int

International Telecommunicaitons Union, WARC-95 Final Acts, in www.itu.int

International Telecommunications Union, WRC-97, Final Acts, in www.itu.int

Intzesiloglou N., Κοινωνία και Νέα Τεχνολογία, (Society and New Technolo-, gy), Sakkoulas, 1992

Iridium LLC, Iridium Reports Fourth Quarter Results, 25 January 1999

Izraelewicz E., Η δικτατορία των αγορών, Ένας ιδανικός αποδιοπομπαίος τράγος (The dictatorship of the markets, An ideal Scapegoat), in Liousis/Ntalis, Οι διεθνείς σχέσεις στη μεταψυχροπολεμική εποχή (International relations in the post cold war era), Exantas, 1999

Jaccard P., Les base et les sources du régime juridique des télécommunications, Lausanne, Diss., 1991

Jachtenfuchs M., and Kohler Koch B., Europaeische Integration, Opladen: Leske and Budrich, 1996

Jacobson H., ITU: A potpourri of Bureaucrats and industrialists, in Cox R., Harold K., Jacobson, M., (eds.) The Anatomy of influence:Decision Making in international Organisation. New Haven: Yale University Press, 1973

Jamalipour A., Low Earth Orbit Satellites for personal communication networks, Artech House Publishers 1998

Joerges Ch., Das Recht im Prozess der Europeaeischen Integration, Ein Plaedoyer fuer die Beachtung des Rechts durch die Politikwissenschaft, EUI Working Paper, 1995

Joerges Ch., The Market without the state? States without a markets? Two essays on the law of the European economy, Florence, EUI, 1996, Working Paper, 96/2

Jones M., (Ed), Cybersociety: Computer-Mediated Communication and Community, 1995

Johnson G., International strategic alliances between Telecommunications Operators, Cambridge: M.Phil.Diss, Judge Institute of Management Studies, 1994

Johnson P., Modern Times: The world from the twenties to the eighties, New York, Praeger, 1983

Johnston R.J., (ed.), Geographies of global change. Remapping the world in the late twentieth century, Oxford, 1995

Kamall S., Telecommunications Policy, Cartermill Publishing, 1996

Kantzenbach E., Scharrer H.-E., Wavermann L., Competition policy in an independent world economy, Baden-Baden: Nomos 1993

Katzenstein P. J., Small states in modern markets, Ithaca, NY, Cornell University Press, 1985

Kazancigil A., (eds) 1986, The state in global perspective, Aldershot: Gower and Paris Unesco

Kebrin S.J., 1995, Beyond Symmetry: State sovereignty in a networked global economy. Department of management, The Wharton School, University of Pennsylvania, Philadelphia June.

Kellner D., Globalisation and postmodern turn, in Globalisation and Europe, Theoretical and Empirical investigations, Roland Axtmann, (ed.), Pinter London and Washington, 1998

Keohane R. O., International institutions and state power, Boudler, CO: Westview press, 1989

Keohane R. and J. S. Nye, 1977, Power and interdependence: World politics in transition, Boston: Little, Brown and Company

Keohane R., Ostrom E. (ed.), Local commons and global interdependence: heterogeinty and cooperation in two domains, London Sage, 1995

Kerever A., Copyright, and space satellites, Congres de la Mer Aigée, 1983

Kiessling T., Blondeel K., 1998, The EU regulatory framework, in Telecommunications, A critical analysis, paper presented to the ITS 98 conference "Beyond Convergence": Communications into the next Millennium", Stockholm, 21-24 June, 1998

Kinderman P., Satellite and WLL: Satisfying the need for Rural Telecoms, Global Communications, Hogan and Cooke, Asia 1998

Klinter W., Satellitenrundfunk und die Problematik des internationalen Urheber- und Leistungsschutzes, Berlin, Schweizer, 1973

Klodt H., Wettbewerb und Regulierung in der Telekommunikation, Tübingen, Mohr 1995

Knauer L. T., Ronald K. M., Thomas M. L., Telecommunications Act Handbook, Rockville/Maryland: Government institutes 1996

Koelsch, 1995, The infomedia Revolution, How it is changing our world and your life

Kosmidis M., Telecommunicaitons Reform in Southern Europe, A comparative analysis of Greece, Spain and Portugal, Diss., EUI, Florence, 1998

KPMG, European Commission, Satellite Communications and their consequences for European telecommunications, Trade and Industry, Main Report, vol.I and II, March 1994, p.19.

Krasner St. D., Power politics, institutions and transnational relations, in Th. Risse-Kappen (ed.), Bringing international relations back in: non-state actors, domestic structures and international institutions, Cambridge University Press, 1995

Kries W. von, Organisation Internationaler Nutzsatellitensystemen, Baden-Baden, Nomos, 1977

Kruse J., Zugang zum Frequenzspecturm, in Ernst-Joachim Mestmäcker (Hrgb) Ein Symposium über Ordnungsprinzipien in Wirschaftsrecht der Telecommunikation und der elektronischen Medien (Kommunikation Ohne Monopole 2), Nomos Baden-Baden, 1995

Kuchling H., Taschnebuch der Physik, 16. Auflage, Leipzig, 1996

Lafferanderie, G., (ed.), Outlook on Space Law in the next thiry Years, The Hague, 1997

Lageman B., Auswirkungen der Globalisierung auf die mittelständische Wirtschaft. Beitrag für einen sammelband des "Roundtable Mittlestand", Essen 1995

La'i'di Z., Τα τείχη της παγκοσμιοποίησης, (The walls of globalisation), in Liousis/Ntalis

Landes D., European expansion, The history of innovation and performance, in Gondolf E., Marcus Irwin M., Dougherty J. P., The Global Economy, Divergent perspectives on economic change, Westview Press, Boulder and London, 1986

La Torre M., Diritto, potere, dominion: Argomenti per una teoria non prescrittivistica del diritto, EUI 1992, Working Paper 92/30

Lazar Fr., Corporate Strategies in Global Markets, in Boyer, Drache, States against markets, 1996

Laudati L., Managing Globalisation: International cooperation in economic regulation, Diss. EUI, Florence, 1998

Lawrence R. Z., Bressant A., Tkatoshi I., A vision for the world economy, The Brookings Institution 1996

Leechor Ch., Harinder S. K. and Sujin H., Structural changes in world industry. A quantitative analysis of recent developments, Washington DC, The world bank technical paper, 1983

Le Goueff Stephan (ed.), Satellite Regulation in Europe, Legal texts and Materials, Kluwer Law International, 2001

Leinwoll, S., From Spark to satellite, New York, Charles Scribner's Son's, 1979
Les dossiers de l'etat du monde, La mondialisation au-delà des mythes, Paris: La Decouverte, 1997

Levinson N. S., Stratégies interorganisationnelles dans les technologies nouvelles: alliances transnationales et mondialisation, in Abdelaziz Mouline (ed.), Les Alliances stratégiques dans les technologies de l'information, 1996

Libois, L.-J., Genèse et croissance des télécommunications, Paris, Mason, 1983

Liousis, Ntalis, Οι διεθνείς σχέσεις στη μεταψυχροπολεμική εποχή (International relations in the post cold war era,), Ekxantas, 1999

Logica, Report, Introduction of S PCS in Europe, Final Report 1995

Long C. D., Telecommunications Law and Practice, 2nd., Ed., London: Sweek and Maxwell, 1995

Long C., Spinks S. O., Alliances, Mergers and Acquisitions in Telecommunications, International Telecoms Review 1997, www.coudert.com/teleacq.htm

Lorance P. and Roos J., Strategic Alliances, formation, implementation and evolution, Oxford, Blackwell, 1992

Lowenfeld Andreas F., International Economic Law, BdII, International Private Investment, 2nd Ed., New York Benden 1982

Luard, E., 1977, International Agencies: The merging framework of interdependencies, Dobbs Ferry, NY: Oceana

Lucatelli A., The making of transnational regimes, Diss., Zürich Universität, 1996

Lyall Fr., The Role of the International Telecommunicaiton Union in Gabriel Lafferanderie

Majid A. A., Legal Status of International institutions, SITA, INMARSAT and EUROCONTROL examined, Dartmouth, 1996

Majone G., Theories of Regulation, in Majone (ed. ), Regulating Europe, 1996, Routledge

Majone J., The rise of statutory regulation in Europe, in Majone, Regulating Europe.

Majone J., Regulating Europe, London: Routledge, 1996

Malanczuk P., Actors: States, International Organisations, Private Entities, in Gabriel Lafferanderie.

Mann F.A., Studies in international law Oxford: Clarendon 1973

Mann M., Is there a society called Euro, in Axtman, Globalisation and Europe, London, Pinter 1998

Mansell Robin, Information, Organisation and Competitivenesss, Networking strategies in the 1990s, in Antonelli C., (ed.), The economics of information networks, Elsevier Publishers, 1992

Mansell R., The new telecommunications, A political economy of network evolution, 1993, Sage publications

Mansell R., Designing Networks to capture policy and regulation issues for the new telecoms environment, in W. Melody (ed.), Telecom reform, principles, policies and regulatory practices, 1997

Marien M., New communications technologies, a survey of impacts and issues, Telecommunications Policy, 20, 5, 1996

Margherio L., Henry D., Cooke S., Montes S., and Hughes K., The emerging digital economy, US department of Commerce, 1998

Markvoort, Hans Government Regulation of Information-Based Product Markets - The case of Computer Software, Diss., St. Gallen Hochschule,1995

Marx K., Freedman R. (ed.), Marx on Economics, Harmondsworth, Penguin, 1971

Marx K., Engels Fr., selected Works, Moscow, Progress, Publishers, 1973

Mattee Mateesco, Space Law, in Encyclopedia of Public International Law, v.11, Law of the Sea, Air and Space, 1989

Matra Marconi, Communications Satellite Systems beyond the year 2000, Report prepared for the European Commission, 1991

Mayntz R., Hughes Th. P., The Development of Large Technical Systems, Campus Verlag, Frankfurt a/M, 1988

Mayes D.(ed.), The evolution of the rules for a single European Market, Luxembourg, 1995

Mazey S. and J.J. Richardson (Ed.) Lobbying in the European Community, Oxford, 1993.

Mc Michael Ph., The new colonialism, in R. Axtman, (ed) Globalisation and Europe: Theoretical and empirical investigation, London, Pinter 1998

Melas K., Παγκοσμιοποίηση, (Globalisation), Eksantas Publications, 1998

Miller, M.J., Vucetic B., and Berry L., (eds), Satellite Communications: Mobile and fixed services, Boston: Kluwer Academic Publishers, 1993

Mittelman J. H. (ed.), Globalisation: critical reflections, Boulder: Lynne Rienner, 1996

Mohammadi D., Winseck J. M., & Boyd B., Media in global context: A reader, New York, Arnold, 1997

Moore, S. F., Law as a Process: an anthropological approach, London: Routledge, 1978

Moran, M., Wright M., The Market and The state, studies in interdependency, MacMillan, 1991.

Mosler Herman, The international society as a legal community, Alphen aan den Rijn: Sijthoff & Noordhoff 1980

Mosteshar S., Development of the Regime for the Low Earth Orbit and the Geostationary Orbit, in Gabriel Lafferanderie.

Müller J., Saadet T., Mobile Communications in Europe, in Steinfielf et al., Telecommunications in transition

Murphy C. and Tooze R., (ed), The new international economy, Boulder, CO: Lynne Rienner publishers International political economy yearbook, volume 6, 1991

Murphy C. 1994, International organisation and industrial change, lobal governance since 1850, New York, Oxford University Press

Murphy C. N., Globalisation and Governance: A historical perspective: in Axtmann (ed.), Globalisation and Europe: Theoretical and empirical investigation, London, Pinter 1998

Naisbitt, J. 1995, Global paradox, New York, Avon Books

Nandi Pr. K. and Shahidullah, Sh. M., (ed.), Globalization and the evolving world society, Leiden; Boston; Koeln; Brill, 1998

Natalicchi G., Telecommunications Policy and integration processes in the European Union, NY, Diss., 1996

NERA, Approaches to the allocation and assignment of the radio spectrum resource, Final Report for the European Commission, November 1997

NERA, Approaches to the Allocation and assignment of the radio spectrum resource, Executive Summary, Final Report for DG XIII, November 1997

Neumann K. H., Interconnection Agreements in Telecommunications, Bad Honnef Wissenschaftliches Institute für Kommunikation, 1993

Nevaer L. E.V., Deck St. A., Strategic Corporate Alliances, New York, Westport/London: Quorum Books 1990.

Nguyen D. G., Competition, Concentration and Competitiveness of the European Manufacturing Industry, in E. M. Noam and G. Pogorel, The Dynamic of Telecommunications Policy in Europe and in the United States, 1994

Nguyen D., Schneider V. and Werle R., Corporate Actor Networks in Telecommunications, Max-Planc Institute für Gesellschaftsordnung, Köln, 93/4

Nihoul P., (ed.), Telecommunications and Broadcasting Networks under the EC law: The protection afforded to consumers and undertakings in the information society, vol. 27, Series of Publications by the Academy of European Law, Trier, 2000

Noam E., Pogorel G., Asymmetric Deregulation: the dynamic of telecommunications policy in Europe and the US, Norwood: Ablex, 1994

Noam E, Wolfson A.J., Globalism and Localism in Telecommunications, Amsterdam, Elsevier, 1997

North D., Institutions: institutional change and economic performance, Cambridge University Press, 1990

Nunnenkamp, P., Gundlach E., Agarwal, J. R., Globalisation of Production and Markets, Tübingen Mohr, 1994

OECD, Towards a global information society: global information infrastructure, global infromation society: policy requirement, Partis, OECD, 1997

Ohmae K., The end of the national state, The rise of regional economies, HarperCollins Publishers, 1995

Ohmae K., The borderline world, power and strategy in the interlinked economy, New York Harper Collins 1991

Ohmae K., The rise of the region-state, Oxford-Clarendon Press,1993

Omnitele, Study on technical and regulatory requirements for open access to Broadband telecommunications networks and services for customers, service providers and content providers, final report, June 1998

Ostry S., and Nelson R. R., Techno-nationalism and Technoglobalism, conflict and Cooperation, The Brookings Institution, Washington DC., 1995

Ostry S., Beyond the border: the new international policy arena, p. 261 ff. in Kantzenbach, Erhard, Scharrer, Hans-Eckart and Waverman Leonard (Ed.), Competition policy in an independent world economy, Baden-Baden: Nomos, 1993

Ovum, LEOs, MEOs and GEOs: The Market Opportunity for Mobile Satellite Services, 1998, p. 225

Palloix Ch., H διεθνοποίηση του κεφαλαίου (The internationalisation of capital), Sakkoulas, 1997

Panitch L., Rethinking the role of the state, in James H. Mittleman, Globalisation, Critical Reflections, 1996

Pelton J. N., Global Talk, The Marriage of the Computer, world communications and man, The Harvester Press,1981

Pelton J. N., Wireless and satellite telecommunications: the technology, the market and the regulation 1995

Pelton J. N., Overview of Satellite Communications, in Encyclopedia of Telecommunications, vl. 14 Marcel Dekkre, 1997

Pentland P. J., (ed), , International theory and European integration, London Faber and Faber, 1973

Petersman, Ernst-Ulrich, constitutional function and constitutional problems of international economic law, , Fribourg/Switzerland, 1991

Petrella R., Globalization and internationalization, in Drache, Boyer, States against Markets, The limits of liberalisation, 1996

Pogorel G. and Noam E. M., Asymmetric deregulation: the dynamics of telecommunications policy in Europe and the US, Norwood: Ablex, 1994.

Posner R., Economic analysis of law , Boston, Little Brown, 1986

Preston L. E., and Duane W., The politics of economic adjustment, international constraints, Distributive conflicts and the state, Princeton, Princeton University Press, 1992

Pribilla P., Telekommunikation und Globales Wettbewerb, in Gebhardt et al.

Rapp L., Satellite Communications and Global Information Infrastructure, in G. Lafferanderie

Razin A., The economics of globalisation, Cambridge University Press 1999

Reasons, Ch. E., Rich R. M., The sociology of law in a conflict perspective, Butterworths, 1978

Reed T., Where will satellites win in the 90s?, in Schorrock D., (ed.) European Communications, Technologies and regulations of the single market, Blenheim on-line publications, 1989

Renaud J.-L., The role of International Telecommunications Union: conflict, resolution and the industrialized countries, in K. Dyson and P. Humphreys (ed.), The political economy of communications , international and European dimensions, , Routledge, London and New York, 1990

Robinson K. G., Bulding a global Information Society, Washington: Aspen Institute 1996

Roobeek A. J. M., Telecommunications: An industry in transitions, in H.W. de Jong (ed) the structure of European industry, 2nd, , Kluver Academy Publishers,1988

Roobeek A. & Broeders J., Telecommunications global restructuring at full speed, in

H.W. de Jong: The structure of the European industry, Kluwer Academy Publishers 1993

Rosecrance R., The rise of the trading state, New York, NY: Basic Books, 1986

Rosenau J. N., The study of global interdependence: Essays on the trnasnationalisation of world affairs, London: Francis Pinter, 1980

Ruigrok W., Tulder R.V., The ideology of interdependence, London, Routledge, 1994

Rutherford Malcolm, Institutions in economics, Cambridge, 1996

Sally R., States and Firms, Mutlintional enterprises in institutional competition, Poutledge, London New York, 1995

Samuels W., Law and Economics: an institutional perspective, The Hague: Nijhoff, 1981

Sanchez B., Integrating technology in a strategic product creation, Champaign: University of Illinois at Urbana-Champaign, 1994, Working Paper

Sassen S., 1996, Losing control? Sovereignty in an Age of globalization, New York: Columbia University Press
Sassen S., The spectial organisation of information industries: implications for the role of the State, in James H. Mittleman, Globalisation, critical reflections, 1996

Satellite Action Plan Regulatory Working Group, Market Access: problems and solutions, Report prepared for the European Commission, October 1999

Sauter W., The relationship between industrial and competition policy under he economic constitution of the European Union, with a case study of telecommunications, Diss., EUI, 1995

Savage J., The politics of international telecommunications regulation, Boulder: Westview P., 1989

Scherer J., Asymmetric re-regulation of telecommunications, in Pogorel G. and Noam E. M., Asymmetric deregulation: the dynamics of telecommunications policy in Europe and the US, Norwood: Ablex, 1994

Scherer, J.(ed), Telekommunikation und Wirtschaftsrecht, Köln: Schmidt, 1988

Scherer J., Telecommunications laws in Europe, 4nd Edition, Butterworth, 1998

Scherer J., Innovationsoffenheit der europäischen Telekommunikations-Regulierung, Konzepte und Instrumente an Beispiel der S PCS-Systeme, Wolfgang Hoffmann-Riem (ed.), Innovation und Telekommunikation, Hans-Bredow-Institut, Hamburg, 2000

Schwindt, R., Marketing I: Marketing Strategy, Management and Research, Chapel Hill: Eno River P, 1995

Scmidt R., Marketing II: Buyer behaviour, industrial marketing, international marketing, Chapel Hill: Eno River P, 1995

Schulte-Braucks E., Europaisches Telekommunikationsrecht in Scherer (ed), Telekommunikation und Wirtschaftsrecht, Köln: Schmidt, 1988

Scott A. M., 1982, The Dynamics of interdependence. Chapel Hill: University of North Carolina Press
Sharkey W., The theory of natural monopoly, Cambridge University Press, 1982

Sharp M., The Community and the New Technologies, in J. Lodge (ed.), The European Community and the Challenge of the future, St Martin's Press New York, 1989

Sharpf F., Marks G., Sckitter Ph., Streeck W., Governance in the European Union, London, Sage, 1996

Siebert, H., (ed.) Wettbewerb und Regulierung in der telekommunication, 1995
Simons and Simons, Telecommunications, The European Union Law, Palladian Law Publishing, 1999

Smith D. A., and Böröcz J. (ed), A new world order? Global transformations in the late twentieth century, Praeger, 1995

Snow M., International Telecommunications Organisations, Economic challenges facing an international organisation, Nomos, Baden-Baden, 1987

Snyder Fr., New direction in European Community Law, Weidenfeld and Nicolson, London, 1990

Snyder Fr., Soft Law and Institutional Practice in the European Union, EUI Working Paper, 1993/5

Snyder Fr., (ed.), European Community Law, vol. I and II, Aldershot, Dartmouth, 1993

Snyder Fr., Interinstitutional Agreements: Forms and Constitutonal Limitations, 1995/4

Snyder Fr., Globalisation and Europeanisation as Friends and Rivals,European Union Law in Global Economic Networks, EUI, Working Paper Law, 1999/8

Snyder Fr., Global Economic Networks and Global Legal Pluralism, Working Paper, Florence, 1999/6

Snyder Fr., Global Economic Networks and Global Legal Pluralism, in Bermann G., Hedeger M., Lindseth P. (ed.), Transatlantic Regulatory Cooperation, Oxford University Press, 2000

Soja E.W., Postmodern Geographies, The reassertion of space in critical social theory, London, 1989

Soroos M., Beyond Sovereignty, The challenge of global policy, University of South Carolina Press, 3rd Edition, 1989

Spulber D. F:, Regulation and Markets, Cambridge, Cambridge, MIT Press,1989

Staffa E. I., Space Technologies in Disaster Management, Global Communications, Hogan and Cooke, Asia 1998

Stauridis S., Mossialos F., Morgan R., Machin H (eds.), New challenges to European Union: Policies and Policy Making, Aldershot: Dartmouth, 1997
Stehmann O., Network Competition for European Telecommunications, Diss, EUI, Florence, 1993

Stopford J., Strange S., Rival States, Rival Firms, Competition for world market shares, Cambridge University Press, 2nd ed., 1994

Strange S., 1988, States and Markets, an introduction to international political economy , New York , Blackwell

Streeck W., Public power beyond the nation state, in Boyer, Drache, States against markets, The limits of liberalisation, 1996

Streeten P. P., Against Minimalism, in Putterman, L., Rueschemeyer D., (ed.), State and market in development, Lynne Rienner Publishers, Boulder & London 1992

Sternberg S., Globalisierung und Lokale Handlungsspielräume I: Globalisierung und
Weltweite Vernetzung, in Gebhardt et. al., Europa in Globalisierungs prozeß von Wirtschaft und Gesellschaft, Franz Steiner Verlag, Stuttgard, 1998

Struzak R., Internet in the Sky: The tests have started...Global Communications, Asia 1998, Hanson Cooke

Stubbs, R. and Geophrey (eds), 1994, Political economy and the changing global order, London, Blackwell

Sussman G. and Lent J. A., (ed), in Transnational communications: writing the third world sage, 1991

Taishoff M., State responsibility and the direct broadcasting satellites, London Pinter, 1987

Tarjanne P., The limits of national sovereignty: issues for the Governance of international telecom, in W. H. Melody (ed.), Telecom reform, principles, policies and regulatory practices, 1997

Taylor, Human Society and the Global Economy, Chapter 4, at http://distanct-ed.bcc.ctc.edu/econ100/ksttext/markets/MARKETS.htm

Teló M., H Ευρωπαικη Ένωση στη μεταψυχροπολεμική εποχή(The European Union in the post cold war era), in Lioushs, Ntalis

Teubner G. (ed.), Global Law without a State, 1997, Aldershot: Datmouth.

Thatcher M., The interaction between the EC and its Member States in Telecommunications Regulation, paper presented in the workshop "The Aftermath of Liberalization: Multi-level Governance and the Regulation of Telecommunications", Florence, 14 November 1997

Thatcher M., The Development of European Regulatory Frameworks: The expansion of European Community Policy-Making in Telecommunications, in S. Stauridis,

E.Mossialos., R.Morgan and H.Machin (eds.), New Challenges to the European Union: Policies and Policy-Making, 1997

Thiebaut J., Technical an legal introduction, functions, status and nationality of satellites, p.10ff., in Telecommunications Satellites and International Law, Colloquium, 1989.

Thieme Ulrich, Rundfunksatelliten und internationales Recht, 1973

Thrift N., A hyperactive world, In Johnston, R.J., (ed): Geographies of global change. Remapping the world in the laet twentieth century, 1995.

Tietmayer H., Die Europäische Wirtschaft zwischen globalen Märkten und internen Herausforderung, Vortrag, EUI, 28/11/96

Tilly Ch., Big Structuers, Large Processes, Huge Comparisons, New York: Russell sage, 1984

UMTS Task Force Report, The road to UMTS, 1996

UMTS Forum, A Regulatory Framework for UMTS, Report no.1, June 1997

United Nations, New York 1999, Department of Economic and Social Affairs, Division for Public Economics and Public Administration, Balancing Market Failure and Government Failure, http://unpan1.un.org/intradoc/groups/public/documents/un/unpan000156.pdf

Ungerer H., Promoting European Communications, The role of the European Community, in D. Shorrock (ed.) European Communications, Technologies and Regulations of the Single Market, Blenheim Online 1989, p. 165

Ungerer H., Comments on Telecommunications Regulatory Reform in the European Community, in Majone G. (eds.), Deregulation or re-regulation: Regulatory reform in Europe and the United States, London: Pinter, 1990.

Ungerer H., 1998, New Priorities for Telecommunications in a competitive world, Balance between competition rules and sector specific regulation, paper presented to the IIC Telecommunications Forum, Brussels, 6-7 July.

US Congress, Office of technology assessment, The World Administrative Radio Conference: Technology and policy implications, Washington, 1993

US Congress, Omnibus Budget Reconcilliation Act of 1993, Conference Report, Title VI, Communications Licensing and Spectrum Allocation Improvement, Report 103-213, August 4, 1993

US Department of Commerce, US Spectrum Management Policy, Agenda for the Future, Washington DC, NTIA Special Publication 91-23, February 1991

Venturini G., Private Actors and Space Law: The Influence of Competition of Satellite Communications, in G. Lafferanderie. (ed), Outlook on Space Law in the next thirty years , The Hague, 1997

Vernon R., Beyond Globalism, Remaking of American Foreign Economic Policy, The free press, NY, 1989

Vernon R., Exploring the global economy, Emerging inssues in Trade and Investment, 1985, Harvard University Center for International Affairs and University Press of Amercia.

Vernon R., The triad as policy makers, Cambridge Centre for Science and International Affairs, 1992

Vernon R., The technology factor in international trade, New York, NBER, 1990

Vernon R., Spar D, Beyond globalism, Boston 1989

Verhoef P., The role of satellite communications policy in the European Community, paper for the EC/EUTELSAT Satellite Seminar, Moscow 18-19 May 1993

Voigt R., Politik und Recht, Beiträge zur Rechtpolitologie, Bochum, 1990

Volney St., Developing and implementing marketing strategies, New York, Praeger, 1986

Wakonen J., Implications of federalism and the principle of subsidiarity in the case of science and technology development in Europe, Turku 1995

Wallace I., The global economic system, London: Unwin Human, 1989

Wallerstein I., The modern world system, new York, Academic 1974

Wang, C-J, Quantitative performance evaluation of satellite networks for global networking-the MVA approach, Collection of Technical Papers, American Institute of Aeronautics and Astronautics (AIAA) 15th international satellite Communications

Waters M., Globalization, London Routledge,1995

Weatherhill St., Law and integration in the European Union, Oxford, Clarendon Press, 1995

Weber M., The theory of social and economic organisation, Talcott Parsons, 1964

Webster F., Theories of the information society, Newbury Park, CA: Sage Publications, 1995

Weinstein O., The telecommunications equipment industry: the great transformation, in Ph. Cooke, E. Moulaert, E. Swyngedouw, O.Weinstein, P. Wells, (ed.) Towards global localisation, 1992, UCL Press

Weiss L., The myth of the powerless state, Cornell, University Press, 1998

Wertheim M., Pearly gates of cyberspace: A history of space from Dante to the internet, Doublday, Sydney: W.W. Norton & Company, 1999

Wickens J. R., Who will be the winners? The making of a top player, Pan Asian PCA 96, January 1996

Williams M., The information society and its implications, in Edward Conor (ed.), The global political economy of communications, Kluwer, 1996

Williamson O. E., Markets and Hierarchies, Analysis and Antitrust implications, 1975, the Free Press

Wiliamson P. J., From barriers to entry, to barriers of survival, Fontainebleau: INSEAD, Working Paper, 1998

Wolfe A., The limits of legitimacy, New York:Basic Books, 1972

Wood M., The globalisation of technology, London Belhaven Press, 1993

Woodward K. (ed.), The Myths of Information: Technology and Postindustrial Culture, London, Routledge & Kegan Paul, 1980

Woolcock St., Hart J. and V. Devren, interdepdencence in the Post-Multilateral Era, Lanham/London 1985

World Wide Satellite Communications, Services and Regulations, Report to clients, introductory volume, September 1991

Wright M., Moran M., The market and the State, studies in interdependence, 1991

Wrona Th., Globalizierung und Strategien der vertikalen Integration, Wiesbaden, Gabler, 1999

Wood L., Network Performance of non geostationary constellations equiped with intersatellite links, Msc Thesis, November 1995, in http://www.ee.surrey.ac.uk/Personal/LWood/Publications

Yip, G., Barriers to entry: a corporate strategy perspective, Lexington Books, 1982

Yoneji M., The information society as a post-industrial society, Institute for hte Information Society, Tokyo, Japan, 1981, p. 29-35

Zdravko M., Tr'cek Fr., On culture and global impact, Territorial cultures and global impacts, in Axtmann R. (ed),Globalization and Europe: theoretical and critical investigation, London, Pinter, 1996

Zhukov G. and Yuri K., International Space Law, New York, 1984

Zuern M., Regierenjenseits des Nationalstaates; Globalisierung und Denationalisierung als Chancen, Frankfurt a.M., 1998, Suhrkamp.

**Articles**

Aamoth, Policy: One law for the rich, Communications International, 1 November 1998

Abramoovitz, Catching up, forging ahead and falling behind, Journal of Economic History, 46:386-406, 1986

Andersen and Kjeil, , European Community Lobbying, European Journal of Political Research, vol. 20, No.2, September, 1991, pp. 173-187

Anderson, Ericsson & Quacomm settle differences and back single 3G standard, Bloomerg news, in www.totaltele.com/secure/view.asp? ArticleID=21784&Pu b=283

Archibugi and Michi, The globalisation of technology, a new taxonomy, Cambridge Journal of Economics, 1995, 19

Arnopoulos, the International Politics of the orbit/spectrum, Annals of Air and Space Law 7, 1982, 215-240

Arroio, Bridging the generation gap, Will the LEOs provide what GEOs couldn't?, Telecommunications Policy, Vol. 21, No.4, 277-282

Aviation Week and Space Technology, Satisfied Demand, Financing Woes Soften Satellite Sector, January 17, 2000, p.151

Backhaus, Piltz, Strategische Allianzen, eine neue Form kooperativen Wettbewerbs, ZFBF, Sonderheft 27 1989, p.1-10

Batty, and Barr, 1994, The electronic frontier: exploring and mapping cyber-space, Futures, 26, (7), 699-712

Baranson, Global Strategic Alliances, why, what, how and where, Multinational Business Studies, 1991, 22, 2.

Baumol, 1986, Productivity growth, convergence and welfare: what the long run data show, American Economic Review, 76:1072-85

Boeke, Surfing among the stars, Via Satellite, November 1997.

Boeke and Robustiano, Global Satellite Survey, Via Satellite, July 1997

Braman, Harmonisation of systems, The third stage of information society, Journal of Communications, 1993, 43(3)

Braman, Horizons of the State, Information Policy and Power, Journal of Communications, Autumn 1995, p. 4 ff.

Breeden, 1993, The globalisation of law and business in the 90s, Wake Forest Law Review, 28, (3)

Brooks, Iridium Settles Freedom to roam, Bloomberg News 25 February 1998

Brown, Global Solution now, IP over Satellites, Via Satellite, October 1998

Bundeep, Inmarsat toe privatize in face of new competitors, Bloomerg News, 25 September 1998

Careless, State of the Satellite Industry, 2001, Via Satellite, December 2001, p. 6 ff.

Carney, Strategy in the telecomunications market, the effects of liberalisation and competition, in Telecommunications Policy, September, 1986

Castells, the net and the self: working notes for a critical theory of information society, Critique of Anthropology, 1996, 16 (1), 9-38

Chioni/Karageorgos, Τηλεπικοινωνίες στην Ευρωπαική Ένωση (Telecommunications in the European Union), Hmerisia, 4 January, 2003.

Christensen, WRC 97: The effect of the satellite, Via Satellite February 1998, p. 80-96

Clark, Mobile and Satellite: Network trials aim to prove 3G's worth, 2 November 1998, in www.totaltele.com/news/view.asp?ArticleID=20484.

Clark, ITU gives up on single 3G standard, in www.totaltele.com/secure /view. asp?ArticleID=21759&Pub=283.

Clark, non immediate solution for troubled Iridium, in total telecom news of 26 April 1999, www.totaltele.com.

Clark, Globalstar Launches European Service, Total Telecom News, 2 february 2000.

Clegg, Kamall, Leung, European Multinational activity in Telecommunications

Services in the Peoples' Republic of China, in Management International Review, issue: 1996/1, pp. 111-137

Codding, The Evolution of the ITU, Telecommunications Policy, v.15, n.4, 1991, 271-285

Colin, The proceduralization of EC law, Telecommunications Policy, vol.22, No.3, pp.243-254

Communications Week International, Spoilt by choice,15 February 1999

Convergence between telecommunications and other media, Telecommunications policy, 22, n.3, 1998

Corbley, Accessing satellite and cellular systems, Dual Mode Handsets to provide the Option, Via Satellite, February 1996

Costelot, The application of competition rules in the telecommunications sector: Strategic Alliances, Competition Policy Newsletter, number 4, v.1, Spring 1995.

Cowhey, The international telecommunications regime: the political roots of regimes for high technology, International organisation, 44, 1990, p.170

Cox, 1992, Mulitlateralism and World order, in Review of international studies 18 161-180

Dales, Beyond the Marketplace, Canadian Journal of Economics, 975, 8, p. 483
de Guiroye, 1993, Community Policy on telecommunications services: The min phases from 1984 to Maastricht, XIII Magasine, No. 11 Office for Official Publications of the European Communicates, Luxembourg: pp.13

Daniel Guy Pie in the Sky?, Communications International, 1/1/ 99

de Vany et al., A property system for the electromagnetic spectrum Stanford Law Review, v.21, n.6, p.1499ff.,

Delcros, Trouchet Didier, Controverse, les ondes appartiennent-elles au domain publique?R.F.D.A., 5, 2, mars-avril, 1989

Demsetz, Barriers to entry, The American Economic Review, March 1982

Detlef, Perspectives on Global communications, Fordham International Law Journal, v. 21, no.2, Dec 1997

Drahos and Joseph, Telecommunications and investment in a great supranational regulatory game, in Telecommunications Policy, 19,8, 1995,

Drucker, 1994, The age of social transformstion, The atlantic monthly, 274, 5

Egami, Satellite link requirements in personal satellite communications, Space Communications, no. 12, 1994

Egan, The role of wireless communications in the global information infrastructure, Telecommunications Policy, vol. 21, no.5, p.3578-385, 1997

Ehlermann, The internal market following the Single European Act, Common Market Law Review, 24, 1987, 361-409

EITO, European Information Technological Observatory 2003, Report, 2003.

Elixmann, Hermann, Strategic Alliances in the telecommunications Service Sector - challenges for corporate strategy, in Communications & Strategies, (IDADE) 1996, p. 57-88

Evagora, CEPT Radio Conferences, Communications Week Internaitonal, July 20, 1992, p.4

Evans, New Satellites for Personal Communications, Scientific American, April 1998

Falvey, 1989, France v. The Commission of the European Economic Community: the power of the Commission to issue a Directive unde rArticle 90. Virginia Law Review, 29, 937-967

Falconer, Fumiyuki, Gudmundson, Time Division Multiple Access Methods for Wireless Personal Communications, IEEE Communications Magasine, January 1995

Farrah, Barbara, Maxwell, Market-based Public Policy, Telephony, June 15, 199

Ferguson, 1992, The mythology about globalisation, European Journal of Communications, 7, 69-93

Fernandez, Internet in the Sky, The Ka-Band Quest Continues, in Via Satellite, March 1997

Fernandez Rob, The year in review, re-defining an industry, Re-defining an Industry, Via Satellite, December 2000, p. 18 ff.

Field, Intelsat at Crossroads, Law and policy , International business 22, 1994-II.

Fleming, US Congress Moves forward on Intelsat Privatization, 19 March 1998, Bloomerg News.

Fleming, US simplifies Sat-Phone approvals, Bloomberg News, 17 December 1998. www.totaltele.com/news/view.asp?articleID=20790&Pub=TT

Foley, Cyberstar strikes strategic alliances in bid to take a lead in satellite data race, Communications Week international, 20 April, 1998

Foley, Mobile & satellite-INMARSAT'S future hangs in balance, in Communications Week international, 2 February 1998

Foley, Eutelsat sets its sights on US market, in Communications Week international, 15 February 1999

Foley, Careful Globalstar to tiptoe into Satellite market. Communications Week international, 20/9/99

Foley, Iridium and Globalstar hit delays, Communications Week International, 21 September 1998.

Foley, Markets lose nerves as satellite ventures stall, Communications Week International, 21 September 1998

Foley, Partners hold the key to global market, Communications Week International, 16/2/98

Foley, Impact felt as high flier Iridium falls to earth, Communications Week International , 24/5/99

Foley, Service Providers reveal sky-high Iridium tariffs, Communications Week International, 2/11/98.

Foley, Iridium and Globalstar hit delays, Communications Week International, 21 September 1998

Foley, Satcoms push into data to meet Net interest, Communications Week International, 16 March 1998

Foley, Eutelsat sets its sights on US market, Communications Week International, 15 February 1999, p. 8

Foley, Cyberstar strikes strategic alliances in bid to take lead in satellite data race, 20 April 1998.

Foley, Cyberstar strikes strategic alliances in bid to take a lead in satellite data race, Communications Week International, 20 April 1998

Foley, Special Report, Satellite Funding- Star attractions, Communications Week International, of 16 August 1999.

Foley, Satellite Funding- StarAttraction, Communications Week International, 16/8/99

Foley, ICO embraces McCaw's new Net strategy, www.totaltele.com/secure /view.asp?ArticleID=25766&pub=CW.

Foley Theresa, Full Service Solutions: Worth a bundle, Via Satellite December 2000, p. 16

Foley, Business Strategies: Satellite companies: gasping for oxygen?, Communications Week International, 16 April 2001

Foley, Satellite Sruge, Communciations Week International, 24 September 2001
Foley Theresa, MSS Survival – Cash and Credibility, Via satellite, June 2002, p. 16.

Forrester, LEOs, GEOs, Meos ; Will they fly?, Via satellite October 1997, p. 26

Forge, The radio Spectrum, Telecommunications Policy, 20, 1, 1996, p. 68 ff.,

Frieden, Satellites in the Wireless Revolution: The need for realistic perspectives, Telecommunications , June 1994

Frieden, Should Inmarsat and Intelsat privatize? Telecommunications Policy, v.18, n.2, 1994, p. 679-686

Frieden, That pesky last mile, Telecommunications Policy, vol. 22, No.2, 1998, pp.133-143

Fuch, ISDN, The telecommunications highway for Europe, Telecommunications Policy November 1992, 635 ff.

Fuchs, 1994, Policy Making in a system of multi-level governance - The Commission of the European Community and the restructuring of the telecommunications sector, Journal of Eureopan Public Policy, 1:2 Autumn, pp. 177-194

Fur, The State between Fragmentation and Globalization, European Journal of International Law, 1997

Garrette Bernard, Quelin Bertrand, An empirical study of hybrid forms of governance structure: the case of the telecommunication equipment industry, in Research policy 23, (1994) 395-412.

Gerber Pat, Teledesic may advance launch of satellite internet project, Totaltelecom news of 28/9/99

Cerny Phillip, 1996, Globalization and other stories: the search of r a new paradigm for internaitonal relations, International jrnal, 51, 617-37

Ghedia Lalji, Smith Keith, Titzer Gary, Satellite PCN – the ICO system, International Journal of Satellite Communications, vol. 17, issue 4, 273-289

Giget, Economices of satelllite communications in the context of itnernational competition, Telecommunications Policy, v.18, n.6 1994, 478-492

Gill, 1992, Economic globalization and the internationalization of authority: limitss and contradictions, Geoforum, 23, 269-283

Gkizeli M., Taaghol P., Evans B.C., Service availability of LEO/MEO mobile satellite systems, International Journal of Satellite Communications, vol. 17, issue 4, p. 291-302.

Gordenker and Weiss, 1995, Pluralising global governance: analytical approaches and dimensions, Third world quaterly, 16, 357-387

Gorove, The geostationary orbit, issues of law and policy, 73 American Journal of International Law, 1979, 444-463

Goueff, Licensing global mobile personal communications by satellite: the quest for Holly Grail? Annals of Air and Space Law, 1997, 22, 1, p. 417-431

Grande, 1994, The new role of the state in telecommunications, in West European Politics, vol. 17 no.3, July, pp. 138-157

Graak, Telecom Operators in the European Union, Telecommunications Policy, vol. 20, n. 5, 1996, pp. 341-355.

Grewlich, Multimedia und globale Kommunication, Wettbewerb und Zusammenarbeit, Europa Archiv 21/1994 S.620-626

Graak C., Telecommunications Operators in the Eureopan Union, Telecommunications Policy, v. 2, n.5, 1996, 341-355, 1.

Goldman Erik, Little LEOs serve an unmet demand: LEO One system architecture optimized to meet market requirements, International Journal of satellite communications, vol. 17, issue 4, 1999, p.225-242.

Grewlich, WATTC-88. Ein Beitrag zum Voelkerrecht der Telekommunikation, RIW 1990, p. 86-100

Grewlich, Positivsummspiel USA-Japan- Europa ?, Aubenpolotik, nr.3, 1988, S. 217-236

Grewlich, ITU, Telekommunikation und Universalitaet, Aubenpolitik, Nr. 4/1989, S. 359-369

Grewelich, WATT-88- Ein Beitrag zum Voelkerrecht der Telekommunikation, in Recht der internationalen Wirtschaft (RIW), 2/1990 S. 86/100

Guy, Pie in the Sky?, Communications International, 1/1/ 99

Haeberle, Gemeineuropaeisches Verfassungsrecht, EuGRZ, 1991, 18, p. 261 ff.also

Haggard and Beth, Theories of international regimes, International organisation, vol.41, no.3, 1987: 419-517

Haris, A Preview of WRC-97, Satellite Communications, May 1997

Hausmann, Joint Ventures, Strategic Alliances and Collaboration in Telecommunications, Regulation, v.14, 1, Winter 1991, p.1

Hawkins, The rise of consortia in the information and communications technology

industries: emerging implications for policy, in Telecommunications policy, 23, 1999, pp. 159-173.

Held, Democracy, the Nation State and the global system, Economy and Society, 20, 2, 138,

Herter, The electromagnetic Spectrum, Natural Resources Journal, 1985, p.25, p.652-663

Hoff Ellen, Introduction, Itnernational Journal of satellite Communications, vol. 17, issue 4, 1999 p. 203-205.

Holmes et al., International competition policy, Telecommunications Policy, vol.20, No. 10, 1996, p.759

Houghton, Pucar, Knox, 1996b, Mapping informaiton technology, Futures, 28(10), 903-917

Hudson, 1979 Implications for Developmencommunications, Journal of Communications. vol. 29 ,no., 1, winter , p. 179-86

Inayatullah and Blaney, 1995, Realizing sovereignty. Review of International studies, 21, 3-20

International Business Week, Telecommunications, the global battle, special report, October 24,1983. p.62

International Telecommunications Union, IMT 2000, the Network aspects, ITU News 8/1998

International Telecommunications Union, ITU Press Release, WRC 97 to shape development of Tomorrow's Radiocommunications Environment, October 23, 1997

Jancso James D., Bruce Kraselsky, The constellation LEO satellite system: a wide-area solution to telecom needs in underserved areas worldwide, International Journal of Satellite Communications, vol. 17, issue 4, 1999, p. 257-271.

Jessop, 1977, Recent theories of capitalist state, Cambridge Journal of Economics 1

Joerges, Taking the Law Seriously, European Law Journal, Vol. 2, July 1996, p. 124

Joynt, The Business: Mobile market set to bed down after mega merger, in Communications Week International, 1 February 1999

Jorde and Teece, Innovation, Cooperation and Antitrust, in Antitrust, Innovation and Competitiveness 55, 1992

Kavanaugh, Star WARCS and new systems, analysis of the US international satellite policy formation, Telecommunications Policy, v.10, n. 2, 1986, p. 93 ff.

Kessler, ISDN Concepts, Facilities and services, Mc Graw Hill Inc., 1990

König, Zeiss,EG-Tlekommunikationsrecht und Lizenzierung von satellitengestützten persönlichen Kommunikationssystemen, EuZW, Heft 5/1999 , p. 133 ff.

König, Anmerkungen: Telekommunikationsrecht im Europa, EuZW, 21/1998, p. 655 ff. EuZW, Heft 21/1998, p. 656

Krasner, 1990/91, Global communications and national power: life on the frontier, in World politics 43 336-366

Kruger, Internet: A SkyBridge too far, Communications International, 1 April 1998

Kruger, Services: back down to earth, Communications International, 1 April 1999

Krugman, Scale economies, Product Differentiation and the Pattern of Trade, American Economic Review 70, 1980, 950,

Krugman, Is free trade passé? Economic Perspectives ,v.1, n.2, Fall 1987, p.131-144.

Kurisaki, Globalisation or regionalization? An observation of current PTO activities, Telecommunications Policy, December 1993, p. 699 ff

Ladeur, Towards a New Theory of Supranationality- The viability of the Network Concept, in European Law Journal, 1997, 33 ff.

Ladeur Rechtliche Regulierng von Informationstechnologien und Standardsetzung, Das Beispiel der Set-Top-Box im digitalen Fernsehen, CR 6, 1999

Lawrence, Patterson, Stuart, Tuck, The Calling Network: a global wireless communications system, International Journal of Satellite Communications, vol.12, 1994, pp. 45-61

Leopold, Low Earth Orbit, Global cellular communications network, Journal of Astronautical Sciences, Vol. 33, April-June 1985, p. 147-161

Lera, Towards multiservice personal communications: perspectives for a sector structure evolution, Telecommunications Policy, vol. 20, n.7, 1996

Levitt 1983, The globalisation of markets. Harvard Business Review (May/June): 92-102

Majone, Cross National sources of Regulatory Policy Making in Europe and in the US, in Journal of Public Policy, 11, 1991

Malanczuk, Ten years of European Telecommunications, Telecommunications & Space Journal, 1994, p.27-51, who sees three main phases in the development of the telecommunications policy in the Community: 1984-1987, 1987-1993 and 1993-today

Maney, Where no billionaire has gone before, Usa Today, Monday, December 23, 1996

Mansell, Strategic Issues in telecommunications, unbundling the information infrastructure, Telecommunications policy, 1994, vol.18, number 8

Marien, New Communications Technology, A survey on impacts and issues, Telecommunications Policy, 20, 5, 1996

Mazur S., A description of current and planned location strategies within the ORBCOMM network, International Journal of Satellite Communications, vol. 17, issue 4, 1999, p. 209-223.

McMichael, 1996, Globalization: myths and realities, Rural Sociology, 61, 25-55

McCaffery Richard, Crowded Orbital Slots test ITU's Influence, Space News, Jan. 27-Feb. 2, 1997.

Mc Millan, Why auction spectrum , Telecommunications policy, 19, n.3, 1995, p. 191-199,

McOmber, Technological autonomy and the three definitions of technology, Journal of Communications, summer 1999, v.49, n.3

Melas, Η αναβίωση της γεωπολιτικής και της γεωοικονομίας. Τα τετράδια της Οικονομίας, (The revival of geopolitics and geoeconomics, The notebooks of Economy), 4-5 September 1999

Menderhausen, 1969, Transnational society v. state sovereignty, in Kyklos 22, 251-273

Michalis, European Union Broadcasting and Telecommunications .Towards a convergent regulatory regime?, European Journal of Communications, v.14, n.2, June 1999

Milgrom, Auctions and Bidding: A Primer, in Journal of Economic Perspectives, Summer 1989, 3, 3-22 Milgrom Paul and Robert J. Weber, A theory of auction and competitive bidding, Econometrica, September 1982a, 50, 1089-122

Miller, Economic regulation and New Technology in the Telecommunications industry, Journal of economic issues, vol. XXX, no. 3 September 1996

Molony, Vendors poised for global network operations role in Communications Week International, 10/5/99

Molony, Consolidation or collision, Communciations Week international, 24 September 2001, Foley, Business Strategies: Satellite companies: gasping for oxygen?, Communications Week International, 16 April 2001.

Monheim, Personal Communications Services: the Wireless Future of Telecommunications, in Federal Communications Law Journal, 1992, vol. 44, n. 2

Monlouis, The future of telecommunications operators alliances, in Telecommunications policy, vol. 22, no.8, 1998, p. 635-641

Mun, Trends in Cordless Communications, in www.cwc.nus.sg/cwcdocs/newsle tter/trends.html

Murray and Mahon Strategic Alliances- Gateway to the New Europe, Long Range Planning 26/1993 S. 102-111

Naftel, The natural death of a natural monopoly: Competition in EC after the Telecommunications Terminal Equipment Judgment, ECLR 1993, S. 105,

Nederveen, Globalization as hybridisation. International sociology, 9, 161-184, 1994

Nelson, 1990, The US tecnology lead: where did it come from and where did it go? Research Policy, 19, 117-132

Nelson, Iridium, from concept to reality, Via Satellite, September 1998, p. 62.

Netti. The state as a conceptual variable, in world politics 20: 559-592

Newman, 1992 Satellite earth stations: new window of opportunity, Financial times, International Telecommunications, October 15, p. 10

Noll, The extraterrestrials are coming, Telecommunications Policy, Vol. 20, No. 2, pp.79-82

Nye and Keohane, Transnational Relations and World politics: an introduction, in International organisation, 25, 1971, Summer no.3, p. 329 ff.

Nye, Intelsat Chief claims no privileges for new skies, in www.totaltele.com/ news/view.asp?articleID=14054&Pub=TT

Nye, ICO looks to Sea and Air for Roaming Customer Base, in www.totaltele. com /news/view.asp?articleID=13050&Pub=TT

Nye, SkyBridge Doubles Capacity fro Broadband Satellite Systems, 1 June 1998, to be found in www.totaltele.com/news/view.asp?articleID016430& Pub= TT

Nye, Olivetti Victory sends warning signals to the European Telecommunications Organisations, total telecom news, 24/5/99, in www.totaltele.com /secure /view.asp?ArticleID=224632Pub0tt&categoryid=0

Nye, Intelsat Drops abominable rules and prepares for market, 8 April 1997, in www.totaltele.com

Nye, ICO Looks to Sea and Air fora Roaming Customer Base, 24 March 1998, www.totoaltele.com

Nye, ICO to Float Mobile Satellite Services, www.totaltele.com news of 18 June 1998

Oberst, Benefits of open markets in satellite communications, Telecommunications Policy, vol. 22, no.9, pp.739-743, 1998

Oberst, New Forms, New Fees, Via Satellite, September 1998, p. 18

Oberst, Spectrum Management, Auctions and More, Via Satellite, October 1998.

Oberst Gerry, The Great Game – Satellite Spectrum Wars, Via Satellite, December 2000

Oberst Gerry, Satellite Fees in Europe – A High Price to Pay, via Satellite, November 2001.

Oh, Global Strategic Alliances in the telecommunications industry, Telecommunications Policy, vol. 20, 9, 1996, p.713-720.

Ohmae, The global logic of strategic allainces, Harvard Business review, März/Apr 1989, S. 143-154

Pandya, Emerging Mobile and Personal Communication Systems, IEEE Communications Magazine, June 1995

Pelaprat Philippe, Inmarsat demarre l'access haut debit mobile sur Thuraya, reseaux no 125/ Fevrier 2003

Pelton, Telecommunications for the 21st Century, Scientific American 1998

Pelton, The Global Satellite Communications Review Panel, Global Satellite Communications Review, Via Satellite, September 1998

Peltzman, 1976, Toward a more general theory or regulation, Journal of Law and Economics, 19, 211-240

Perucci and Cimatoribus, Competition convergence and asymmetry in telecommunications regulation, in Telecommunications Policy, v.21, n.6, 1997.

Pfeifenberger and Houthakker, Competition to international satellite communications services, in Information economics and Policy, 10 , 1998, 403-430

Public Network Europe (PNE), Mobile via Satellite: bent pipe dreams ?October 1997

PNE, UMTS auctions: a bid for the future? July/August 1998, p.50

PNE, Satcomms: tortoises and hares? July/August 1998, p. 25

PNE, Mobile via satellite: bent pipe dreams? PNE, December 1997, p.45

PNE, Getting lost on the road to the information society, October 1999

PNE, Cellular star burning brightly, Public Network Europe, February 1999, p.38

PNE, UMTS auctions: a bid for the future, in, July/August 1998, p. 49

PNE, The haste of a new generation?, December 1999/ January 2000. p. 26 ff.

PNE, Celestial ruination, October 1999

PNE, Flying high, laid low, November, 1999, p. 21

PNE, Of course they are big, but what are they there for?, November, 1999

PNE, Telecomms investment- caveat emptor, July / August 1999

PNE, ISOs: spun out of orbit, June 1999, p.28

PNE, Telecomms Investment-caveat emptor, July/August 1999., p.21ff.

PNE, Stop the world, I want to get off, September 1999

PNE, GMPCS: Now the talking can start, October 1998

PNE, Satcomms: nervous systems?, November 1998

PNE, Getting lost on the road to the info society, October 1999

PNE, In Cyberspace no one can hear screeaaammming, October 1999

Rieger Elmar, Leibfried Stephan, Globalisation and the Western Welfare State, ZeS, 1995

RSDL, Globalstar opens services to US consumers, in www.totaltele.com /secure/view.asp?articleID=25308&Pub=TTcategoryid=625

Rothblatt, ITU Regulation of satellite Communication, 19 Standford Journal of International LAW, 1, 7, 1982.

Rusch, Perspective : Investors should take heed of satellite's past, Communications Week international , 4, October 1999

Samers, Migration and the spatial rent, in Review of international political economy, Summer 1999,

Satellite International, EMSAT unveiled, Satellite international, volume 1, number 17, October 21, 1998,

Satellite Today Via Online, 1998 Global Satellite Survey, www.satellitetoday. com/viaonline/survey

Satellite Today, Market Snapshot, in www.satellitetoday.com/snapshot /index. html November 22, 1999

Satellite Week, Iridium delays paging service until Mid-November, vol., 20 no 44, 2 November 1998

Sauter, The telecommunications law of the European Union, European Law Journal 1,1, pp.92-111, 1993

Schamp 1996, Globalisierung von Produktionsnetzen und Standortsystemen. In : Geographiche Zeitschrift, 84, s. 205-219

Scharp, Community and autonomy: multilevel policy-making in the European Union, Journal of European Public Policy, 1994, pp.219 ff.

Scherer, European Telecommunications Law: The framework of the Treaty, European Law Journal, 1987, p. 354 ff

Schertler, Richard, Zernic, ACTS' legacy: The Ka-Band explosion, Aerospace America, February 1996

Schmoch and Schöring, Technological Strategies of telecommunications equipment and manufacturers, a patent analysis, Telecommunications Policy, 18, 5, 1994,

Sendra, Strategic Alliances for Innovation, A Comparative study of the relationship between innovation and the patent/antitrustmechanisms of the United States and the European Economic Community, International Tax and Business Lawyer, 9, (2), 1992

Shetty, Cover story: on a different wavelength, Communications international, 1 May 1999

Shetty, Communications International, ITU Plenipot 98 plays Paper satellite Ping-Pang, 29 October 1998

Shaw, 1997, The state of globalisation: toward a theory of state transformation, Review of International Political Economy, 4, 3

Slot, Case studies: C-202/88, c-41/90. C-260/89, Common Market Law Review, 28, 1991

Snyder, Governing Economic Globalisation : Global Legal Pluralism and European Law, European Law Journal, vol. 5, (4), December 1999, p. 334-375

Space Bus. News, ITU Still Awash in Paper Filings, LEO Constellations Adding Confusion, October 30, 1996

Spicer, Responses to globalization of telecommunications, Telecommunications policy, December 1992, p. 729 ff.

Spiegel, Let the big boys go first, Forbes Magasine, October 19, 1998.

Straubel, Telecommunications Satellites and Market Forces: Should the Geostationary Orbit be regulated by the FCC?, North Carolina Journal of International law and commercial regulation, Winter 1992, p. 215

Stuliadou, the Application of Competition Rules in the Telecommunications Sector, Telecommunications Policy, 21(1), 1997

Sung, WARC-92: setting the agenda for the future, in Telecommunications Policy, November 1992, p. 624 ff.

Sussman, 1995, Transnational communications and the dependent integrated stae, Juornal of Communications, 45 (4)

Sweet, Judicialization and the construction of governance, in Comparative Political Studies, April 1999.

Taaffe, Two-way satellite operators aim at SMEs, Communications Week International, 6 May 2002.

Talens, Satellite Co-ordination ,An introduction, Via Satellite, p.92, April 1998

Taylor, 1993, Satellites Technology: more vital than ever, Financial times, International Telecommunications, October 18, p. 12

Tedeschi, et al., Power, Influence and Behavioural Compliance, 4 Law and Society Review, p. 21, 1970

The Economist, The final frontier, July 27, 1996

The Futurist, Getting set for the coming millenium, 29, 2, 1995,

The Guardian, Saudi billionaire makes $200m bet, April 15, 1998

The return of Iridium , in http://www/forbes.com/2001/11/30/1130tentech.html

Thompson, The problem f globalisation: International economic relations, national economic managements and the formation of trading blocks, Economy and Society, vol.21, No.4, November 1992, p. 358

Thompson, 1992, The evolution of managed economy in Europe, in Economy and society, 21, 2, 129-151

Tisanes F., 1988, Technical cooperation and agreements between firms, Science Technology Industry Review, no., 57-119, 1992

Titch, Is PCS more than just a portable pocket phone? Telephony, June 17, 1991

Totaltelecom News, Intelsat launches New skies at private sector, in www.total tele.com/news/view.asp?articleID=13635&Pub=TT

Trubek, Dezalay, Buxhanan, Davis, Global restructuing and the law: Studies of the internationalisation of legal fields and the creation of international arenas, Case Western Reserve Law Review, 1994

Tucker, Action Plan for mobile phones, Financial Times, 24 November, 1994

Tyler, Bednarczyk, Regulatory institutions and processes in telecommunications , in telecommunications policy 1993, S. 323-342

Ungerer, European Policies and Regulation, Telecommunications Policy, December 1992, p.715 ff

Verhoef, 1993, Satellite Communications, XII Magazine, No. 11, Office for Official Publications of the European Communities, Luxembourg, pp.25-32

Via Satellite, Intelsat in the 90s: shaping a global society , Via Satellite, April 1993

Via Satellite, The Early Bird catches the World, April 1997, p. 56

Villeneuve and Kaufmann, Creating successful technology-based corporate partnering arrangements, 9 Computer Law, 10, 1992

Virginia Journal of International Law Privatization of Satellite Cooperatives: smothering a Golden Goose, vol. 36, p. 1001 ff, 1996

von Baggehufwudt, Dienstleistungsmonopole in der Telekommunikation unter EG-rechtlichen Aspekten, Archiv PT 1993, p. 174-179

Walker, Some circular orbit patterns providing continuous whole earth coverage, Journal of the British Interplay Society, vol. 24, pp. 369-384, 1971

Warner, Wireless technologies, creating competition in the local exchange market, in Common Law Conspectus, Winter 96, vol. 4, pages 51-78

Warnery, 1993, Telecommunications Services: Towards Liberalisation, XIII Magazine, No.11, Office for Official Publications of the European Communities, Luxembourg, pp. 6-10

Warwick, Mobile satellite: up in smoke, Communications international, 1 October 1999

Weidenbaum, When Business cross internaitonal borders, Transnational Lawyer, vol.7, 1994, p.129-132

Whightman, 1987, Competitive Avantage throuh information technology, Journal of General Management, 12, 4

Wiessner, the public orgder of the geostationary orbit: Blueprints for the future, 9 Yale J. World Public Order,217, 218-20 (1983)

Wilson, The International Telecommunications Union and the Geostationary Satellite Orbit: an overview , Annals of Air and Space Law 23 (1998): 241-270

Wireless Europe, Cyberstar introduce broadband services, 29/10/98

Wong, The paper "satellite" chase: the ITU prepares for its final exam in Resolution 18, The Journal of Air Law and Commerce, 63 (1998), 4, 849-879

Wold Robert N., Europe's Pursuit of Internet and Broadband Revenues, Via Satellite September 2001, p.37

Wolfurm (The principle of Common Heritage of the Mankind, ZaöRV, 1983, 312

Wright and Taylor, Policy and Regulatory Framework for the mobile satellite services, Telecommunications Policy, 1996, 208, 549-555

Wright, The European Commission and the satellite industry combine forces to open markets, Telecommunications Policy, 1998 Vol. 22, no. 4/5, pp.267-272.

Wright, sovereignty and universal service: sacred cows in the satellite industry, telecommunications policy, 23, 1999

Wriston, Technology and Sovereignty, in Foreign Affairs, Nr. 89/1988, p. 63-75

Xavier, The licensing of telecommunications suppliers, Beyond the EU's directive, Telecommunications Policy, v.22, n.6, 1998

Ypsilanti and Xavier, Towards next generation regulation, Telecommunications Policy, 1998, vol. 222, no. 8, p. 645

**European Community Documents**

Commission Communication on the introduction of satellite personal communications, COM(93)171

Commission, Com(93)652-COD 487, Proposal for a European Parliament and a Council directive on a policy for the mutual recognition of licenses and other national authorisation for the provision of satellite network services, and/or satellite communications services, aiming at setting a framework for the mutual recognition of licenses within the EU, OJ L 36/1, 1994

Commission Communication on the Consultation on the Green Paper on mobile and personal communications in the European Union, COM(95)158 final. 03.05.95

Commission Communication, Towards a European Numbering Environment: Green Paper on a numbering policy for telecommunications services in Europe, COM (96) 590

Commission Communication to the Council, the European Parliament, the Economic and Social Committee and the Committee of the Regions on the further development of mobile and wireless communications. Challenges and Choices for the European Union, COM(97)217

Commisison Communication to the European Parliament and Council on the World Radiocommunication Conference 1997, COM (97)304

Commission Communication to the Council, the European Parliament, the Economic and Social Committee and the Committee of the Regions: Strategy and policy orientations with regard to the further development of mobile and wireless communications (UMTS). Outcome of the public consultation and proposals for creating a favourable environment. COM(97)513, 15.10.97

Commission Communication to the Council, the European Parliament, the Economic and Social Committee and the Committee of the Regions, ECU Action Plan: Satellite Communications in the Information Society, COM (97) 91 final, 05.03.1997

Commission Proposal for a European Parliament and Counicl Directive on a coordinated introduction of mobile and wireless communications (UMTS) in the Community, COM(98)58

Commission Communications from the commission to the European Parliament and Council on Radio frequency requirements for Community policies in the context of the World Radiocommunications Conference 1999, COM (1998)298

Commission Communication to the European Parliament, the Council, the Economic and Social Committee and the Committee of the Regions, Fifth Report on the Implementation of the Telecommunications Regulatory Package, COM(99) 537, 10.11.99

Commission Communication to the Council, the European Parliament, the Economic and Social Committee and the Committee of the Regions, Electronic Communications: the road to the knowledge economy, COM (2003)65, 11.2. 2003

Commission Directive of 16 May 1988 on Competition in the markets in telecommunications terminal equipment 88/301/EEC, OJ L No 131/73, 27 May 1988

Commission Directive of 28 June 1990 on Competition in the markets for telecommunications services, 90/338/EEC, OJ L No 192/10

Commission Directive of 13 October 1994 amending Directive 88/301/EEC and Directive 90/388/EEC in particualr with regard to satellite communications, 94/46/EC, OJ No L268/15, 19.10.94

Commission Green Paper on the Development of the Common Market for telecommunications services and equipment, COM(87)290, 30.06.87

Commission, Directive of 16 January 1996 amending Directive 90/388/EEC with regard to mobile and personal communications 96/2/EC, OJ L 20/59, 26.01.96

Commission Green Paper on a Common approach in the filed of satellite communications in the European Community, COM(90)490, 20.11. 90

Commission Green Paper on a Common approach to mobile and personal communications in the European Union, COM (94)145 final, 27.04.94

Commission, Green Paper on the liberalisation of telecommunications infrastructure and cable TV networks- Part II: A common approach to the provision of infrastructure in the European Union, COM(94) 682

Commission Green Paper on the Convergence of the telecommunications, media and information technology sectors, COM(97)623

Comission Green Paper on Radio spectrum policy in the context of the European Community policies such has telecommunications, broadcasting , transport , research and development COM(98)596

Commission Guidelines on market analysis and the assessment of significant market power under the Community regulatory framework for electronic communications networks and services (2002/C 165/03)

Commission Proposal for a Council Decision on the Measures of the Measures of the Member States regarding the area of radio freqeuncies, COM(93)

Commission, Proposal for a Council Directive on policy for the mutual recognition of licences and another national authorisation for the provision of satellite network service and /or satellite communications services, COM(93) 652

Commission, Proposal for a Parliament and Council Decision on an action at a Union level in the filed of satellite personal communications, COM(95)529

Commission Proposal for a European Parliament and Council Directive on a common framework for general authorization and individual licences in the field of telecommunications services, COM (95) 545

Commission Report on European Union Telecommunications Policy, DG13, March 1999.

Commission, The impact of satellite development of R&D, Report 30 April 1998

Commission, The EU Action Plan: Satellite Communications in the Information Society, Recent activities, present situation and outlook, Staff Wokring Paper, SEC(1999)250, Brussels, 16. 2.99.

Council Directive of 24 June 1986 on the initial stage of the mutual recognition of type approval for telecommunications terminal equipment, 86/361/EEC, OJ No L 217/21, 05.08.86

Council Directive of 25 June 1987 on the frequency bands to be reserved for the coordinated introduction of public pan-European cellular digital land-based mobile communications in the European Community, 87/372/EEC OJ L No 196/85, 17.07.87

Council Directive of 28 June 1990 on the establishment of the internal market for telecommunications services through the implementation of the open network provision, 90/387, OJ L 192/1, 24.07.90.

Council Directive on the introduction of a pan-European land-based public radio paging in the Community, 90/544/EEC OJ L No 310/28, 9.11.90

Council Directive of 29 April 1991 on the approximation of the laws of the Member States concerning telecommunications terminal equipment, including the mutual recognition of their conformity, 91/263/EEC; OJ L No 128/1, 23.05.91

Council Directive of 3 June 1991 on the frequency band to be designated for the coordinated introduction of digital European cordless telecommunications (DECT) into the Community, 91/287/EEC OJ L No 144/45 of 08.06.91

Council Directive 93/97/EEC of 29 October 1993 supplementing Directive 91/263 in respect of satellite earth station equipment, OJ No L 290/1, 24.11.93

Council Recommendation of 25 June 1987 on the coordinated introduction of public pan-European cellular digital land-based mobile communications in the Community, 87/371/EEC, OJ L 196/81, 17.07.1987

Council Recommendation of 3 June 1991 on the coordinated introduction of digital European cordless telecommunications (DECT) in the Community 91/288/EEC, OJ L No 144/47, 08.06.1991

Council Recommendation of 9 October 1990 on the coordinated introduction of pan-European land-.based public band-based public radio paging in the Community, 90/543/EEC, OJ No L 310/23, 09.11.90

Council Resolution of 28 June 1990 on the strengthening of the European-wide co-operation on radio frequencies, in particular with regard to services with a pan-European dimension, OJ C No 166/4, 07.07.1990

Council Resolution of 14 December 1990 on the final stage of the coordinated introduction of pan-European land based public digital mobile cellular communications in the Community, OJ C No 329/25, 31.12.1990

Council Resolution, of December 1991 on the development of the common market for satellite communications services and equipment, OJ C No 8/01, 14.1.92

Council Resolution of 7 December 1993 on the introduction of satellite personal communications services in the Community, OJ C No 339, 1, 16.12.93

Council Resolution of 22 December 1994 on the further development of the Community's satellite communications policy, especially with regard to the provision of, and access to space segment capacity, 1994, OJ C No 379, 4

Decision No 710/97/EC of the European Parliament and the Council of 24 March 1997 on a co-ordinated authorisation approach in the filed of satellite personal-communications services in the Community, OJ L No 105/4, 23.04. 1997

Decision No 128/199/EC of the European Parliament and of the Council of 14 December 1998 on the introduction of a third-generation mobile and wireless communications system (UMTS) in the Community, L No 17/1 22.1.1999

Directive 97/13/EC of the European Parliament and of the Council of 10 April 1997 on a common framework for general authorisations and individual licenses in the filed of telecommunications services, OJ L No 117/15, 07.05.97

Directive 97/33/EC of the European Parliament and of the Council of 30 June 1997 on interconnection in Telecommunications with regard to ensuring universal service and interoperability through application of the principles of the Open Network Provision (ONP), OJ L No 199/33, 26.7.99

Directive 98/13 of the European Parliament and of the Council of 12 February 1998 relating to telecommunications terminal equipment and satellite earth station equipment, including the mutual recognition of their conformity, OJ L No 74/01 of 12.03.98

Directive 98/10 of the European Parliament and of the Council of 26 February 1998 on the application of open network provision (ONP) to voice telephony and on universal service for telecommunications in a competitive environment, OJ L No L 101/25, 1.4.98

Directive of 9/3/99 of the European Parliament and the Council on Radio Equipment and Telecommunications Terminal Equipment and the Mutual Recognition of their Conformity, 1999/5/EC, OJ L No 91/10, 7/4/99

**CEPT/ERC/ECTRA Documents**

CEPT, CEPT Vade Mecum, September 1994

ECTRA/DEC/(97)02 of 3 July 1997on harmonisation of authorisation conditions and co-ordination of procedures in the filed of S-PCS in Europe, operating within the 1.6-2.4 and 1.9/2.1 bands

ERC/DEC/(97)03 of 30 June 1997 on harmonisation of authorisation conditions and co-ordination of procedures in the filed of S-PCS in Europe, operating in the frequency bands 1.610--1626.5 Mhz, 2.483,5-2.500 Mhz, 1980-2010 Mhz, 2170-2200 Mhz

ERC/DEC/(97)04, ERC Decision of 30 June 1997 on transnational arrangements for the Fixed services and the mobile satellite services in the bands 1980-2010 MHz and 2170-2200 in order to facilitate the harmonized introduction and development of the satellite personal communication systems

ERC/DEC/(97)05, ERC Decision of 30 June 1997 of free circulation, use and licensing of Mobile Earth Stations of Satellite Personal Communications Services operating in the bands 1610-1626.5 MHz, 2483,5-2500, 1980-2010, 2170-2200 MHz within the CEPT

ERC Decision of 10 March 1999 on free Circulation, Use and Exemption from individual Licensing of Mobile Earth Stations of S PCS <1GHz systems, ERC/DEC/(99)05

ERC/DEC/(99)06, ERC Decision of 10 March 1999 on harmonised introduction of S PCS operating in the bands below 1 GHz